THE FUTURE OF NOW

未来即现在

2014国际青年设计师邀请展作品集
INTERNATIONAL
YOUTH DESIGN EXHIBITION WORKS

刘元风 主编

"中国传统服饰文化的抢救传承与设计创新"博士人才培养项目
2011计划——中国民族艺术传承与传播协同创新中心
教师队伍建设——创新团队项目（民族服饰文化传承与设计创新团队）

中国纺织出版社

主办：	Host：
北京服装学院	Beijing Institute of Fashion Technology
承办：	Organizer：
服装艺术与工程学院	School of Fashion
艺术设计学院	School of Art and Design
造型艺术系	The Department of Plastic Arts
北服创新园	BIFTPARK
民族服饰博物馆	Museum of Ethnic Costumes
协办：	Co-organizer:
北京国际设计周	Beijing Design Week
中国服装设计师协会	China Fashion Association
中国纺织服装教育学会	China Textile and Apparel Education Society
中央美术学院	China Central Academy of Fine Arts
三星电子	Samsung Electronics
苹果公司	Apple Inc.
今日美术馆	Today Art Museum
时尚芭莎	Harper's Bazaar
IFFTI (International Foundation of Fashion Technology Institutes)	IFFTI (International Foundation of Fashion Technology Institutes)
IBIC	IBIC
有个微电影工作室	Younger Micro-film Studio
支持：	Support：
《艺术设计研究》杂志	*Art and Design Research* Magazine
《装饰》杂志	*ZhuangShi* Magazine
《设计管理》杂志	*Design Management* Magazine
摄影：	Photography:
陈大公	Dagong Chen
书籍设计：	Graphic Design:
李煌	Huang Li

前言
FOREWORD

现在是新时代的清晨。

由已知和未知、已然与未然交织的现在，是希望的源泉，是达成理想的原因。现在既是收获也是播种，是果实也是种子。有创造性、有使命感的现在，是我们信任明天的理由。作为一所卓越的时尚学府，在她的纪念日，我们为您呈现了另一番景象。

来自世界 21 个国家和地区的 70 余位青年设计师，联袂展现了他们各自的才华。这些源自不同文化背景、基于不同观念、设计形态各异的作品，既构成了交相辉映的丰富性，也传达了一种共同的意愿：设计不仅改变人的情趣和生存方式，更应该用心建设理想的现在。

保持了创造热度的青年设计师们，是设计界的清晨。沐浴着熠熠晨光，足以令人确信，每一个新的时代，都是被一个个有创造力的人一点一点地展开。

现在就是这样决定着未来……

刘元风
北京服装学院院长

Now is the morning of a new era. Intertwined with the known and unknown, existent and nonexistent, is the source of hope, and the reason for pursuing. Now is the time for harvesting and for sowing, it's the fruit and also the seed. Being in this creative and flourishing moment, is the reason to be confident in tomorrow. On the anniversary of this outstanding institution in fashion, we, present you another extraordinary morning.

70 young designers from 21 countries and territories are here to demonstrate their individual talents. These works are born from different cultural background and crafted by diverse concepts and design methodologies. Presented together in their own amazing beauties, these creations share one common desire: Design will not only change the lifestyle for the future, but also construct every ideal moment of the present.

This group of talented young designers, is the morning of this design world. Stay in the youth, it is certain that each of the new era is gradually opened by every single inspired mind.

Yuanfeng Liu
President
Beijing Institute of Fashion

目录
CONTENTS

| 一 服装设计作品 | 008
FASHION DESIGN

二 艺术设计作品 | 074
ART AND DESIGN

三 造型艺术作品 | 120
PLASTIC ART DESIGN

四 创意说 | 132
IDEAS

附录:
设计师
DESIGNERS

服装设计作品
FASHION DESIGN

服装服饰静态展区展示了来自13个国家及地区（包含10个APEC成员）的29位设计师的76套服装服饰作品，其中女装作品57套，男装作品19套。这些优秀作品从多角度展示了国际青年设计师的服装设计创新能力，从中可以感受到来自不同文化传统的国度对服装服饰设计的理解与诠释。

The Fashion Static Exhibition displays 76 sets of clothing works of 29 designers from 13 countries and regions (containing 10 APEC members), among which are 57 sets of women's wear and 19 sets of men's wear. Those outstanding works of international young designers show their creativity from different angles, from which the audience may gain various understanding of the design and interpretations from different cultural traditions of different countries.

"软-雕塑"。该雕塑是由天然、细粒材料雕刻而成的男性身体。借助材质展示雕塑的外部服装。通过对面料材质的重塑、破坏与雕刻,重新塑造整个结构与外形。

It is a celebration of masculine body forming by the raw and fine material. All the textiles turn into the media for interpreting the external apparels of the body. There is via of folding, breaking and sculpting the fabric to recreate the structures and surface.

创作灵感来源于一位自信的个人主义女性形象,是三款独一无二的婚纱。通过使用皮革、毛皮、饰边以及其他轻型面料转变传统方式。所使用的材质缔造出一种令人充满幻想的外观,并且几乎所有装饰细节都是纯手工制作的。形成对比的轮廓也营造出之中的正能量。作品使用了大量的白色与金属颜色,搭配粉色与波尔多色调。

The collection inspired by a self-confident individualistic woman consists of three distinctive pieces of one-of-a-kind wedding gowns. It builds on antidotes to a traditional approach by introducing the symbiosis of leather, fur, lace and other lightweight fabrics. The materials used to create an airy look with ornamentals details mostly created by hand. The contrast of silhouettes creates a stream of positive energy. The collection employs the scale of white and metallic colors with an accent of pink and Bordeaux tones.

灵感来源于到巴黎后开始的新生活,这种状态使我更加轻松地去观察生活的变化:街边的垃圾桶、路的形态、生活的用品、每日乘坐地铁和公交都看到的公共设施、亮丽和让人w觉得新鲜的色彩。喜欢能给生活带来方便的东西,于是关注于每件衣服上那个最具功能性的口袋。服装的廓型和色彩都来自于日常生活,想要设计出轻松且能带给人愉悦心情的成衣系列。

Design specification: The inspiration comes from my new life after arriving at Paris, which helps me observe those changes in life, trash cans along the street, forms of the road and articles of daily use with a carefree mind. Besides, I also observe the facilities while taking subway and bus, those bright colors which enable people feel fresh. I love things that can facilitate our life, therefore I focus on designing the most functional pocket on each clothes. The silhouette and color of clothing are all derived from our daily life, so I want to design ready-to-wear collections that can keep people in a cheerful mood.

狄梦洁所设计的男装"卷",灵感来源于赫迪·苏莱曼摄影日记,展现了巴黎、伦敦及纽约目前以及未来作为时尚之都的景观。每张照片都通过狄梦洁的设计阐释了一个故事,从开始阶段的"困惑"到"发现"再到"迸发"的一系列过程。透过每个形象的阐释,狄梦洁根据信息大爆炸的当今时代设计了不同的作品。人类身处一个充盈着信息的世界,卷入无限的循环中。

Di Mengjie`s Menswear collection "Entangled" was inspired by Hedi Slimane Photographic Diary which released his vision of Paris, London and NYC as fashion cities of now and future. Each his photo symbolizes a story through Mengjie`s design.

TIMEMACHINE G 品牌 2014 年秋冬系列。设计灵感来源于对 20 世纪 60 年代中国的服饰研究。提取了当时服装色彩和样式上的特点，结合英国西装高级定制的技术与细节，将设计融入于服装细节当中。其中包括丰富的设计元素。从色彩、廓型设计到剪裁都体现了中西结合的特点。

Brand TIMEMACHINE G, series of autumn and winter 2014. I am inspired by the costume studies of China in 1960s. I extract the characteristics in terms of costume color and style, combine the skill and detail of advanced customization of British suits, and combin the design with details. It contains abundant design elements, and reflects the characteristics of the integration of Chinese and western from color, profile design and clipping.

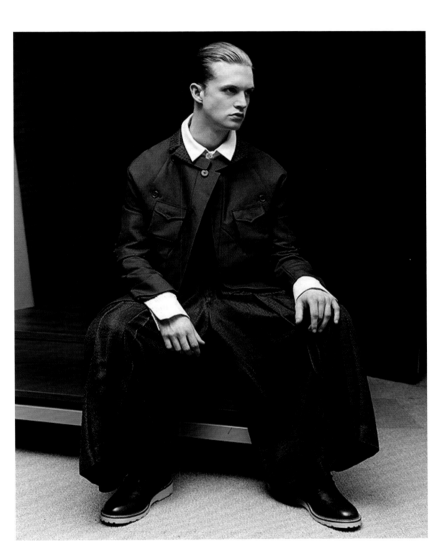

保罗·兰德曾经说过：设计既是一个动词，又是一个名词；是开始，同时也是结束；是想象的过程，也是想象的产物。
系列灵感来自对与生俱来的图形——自身胎记的解读。对于胎记的衍生和发散，妙趣在于过程。在实现意图的过程中获得感动，任何人都能以与作者相同的视点去追寻，以此共享这种价值观。这些实际上就是存在于身边的现实。

Paul Rand once said that: design is a verb, but also is a noun; a start also an end; the process of imagination, also the result of imagination. It is inspired by the interpretation of the within-born pattern—the birthmark. The interesting point of the derivation and expansion of birthmark lies in the process. Being touched on the process of achieving intention and pursuing in the same perspective as the designer, and sharing this kind of value, all these are the realities which exist around us.

繁忙而单调的生活让一些都市人产生了希望回归自然、放松身心的追求。本设计作品利用纱质面料的轻盈质感和半透明的特性，用布料透叠的手法产生深浅变换的视觉效果，营造出自然界山水云相互交映的意象。

The busy and toneless life drives people to pursue returning to nature and relaxing. This design uses the ethereal texture of yarn material and the character of semitransparent texture to create an image of the mountain, water and cloud reflecting each other, and the design also uses overlapping transparent materials to create a visual effect which changes from deep to light.

本系列服装命名为《紫禁遗梦》，其设计灵感来源于电影《末代皇帝》的序曲。作为中国封建社会的最后一位皇帝，溥仪一生跌宕坎坷，然而，他却拥有一段快乐安逸的孩童时光。以紫禁城的各个职能部门为划分基础，本系列的每套服装都拥有各自的主题：弘仪阁的银库管理员、内务府的鸟官、宫廷戏班的青衣、兵工厂的兵器工程师、钦天监的监正、针工局的衣女……作者将古典宫廷生活与当代艺术和前卫相结合，穿越于未来与现在，意在把本系列服装设计建设成为一个展馆，将紫禁城内的神秘以另类的方式展现出来。

The name of the series clothes is "Leftover Dream of the Forbidden City", and the inspiration comes from the overture of the film The Last Emperor. As the last emperor in China's feudal society, Henry Pu Yi's life was filled with ups and downs, but he experienced a happy childhood. Each clothes, based on the functional departments of the Forbidden City, has its own theme: treasury administrator of Hongyi Pavilion, bird officer of the imperial household department, female roles of the theatrical troupe, weapon engineers of the arsenal, imperial astronomers, clothing maids of needlework bureau, etc. The designer combines the classical palace life with modern art and Avant-garde, running across the future and the present, with the aim to construct the series clothing into an exhibition hall, and reveal the mysterious inside the Forbidden City to people in a special way.

设计灵感来源于对中国传统旗袍的印象,采用毛毡、绢与戳绣的结合,以简洁、明快的廓型与线条强调正、负型之间所呈现的独特美感。

It is inspired by the impression of Chinese traditional cheongsam, which combines wool felt, raw silk and embroidery, stressing the unique beauty presented between positive and negative through simple and vivid silhouette and line.

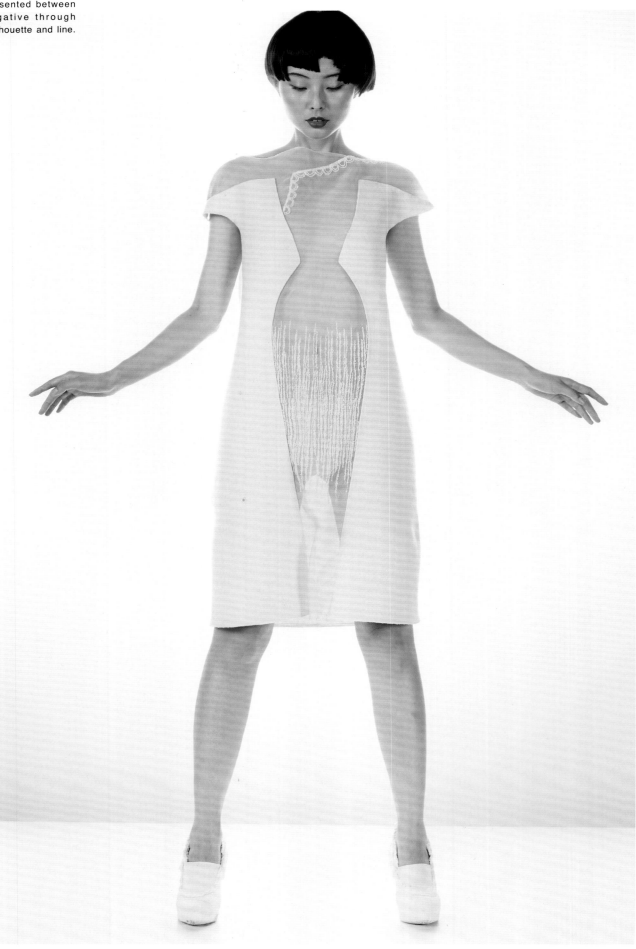

我为那些有趣的女性设计，她们与众不同，追求新挑战、感激生活、向往自由与情感。

我的设计灵感来源于各种理念：遥远未来的人类生活、脑神经元的发展情况、人类情感的脆弱与力量。设计的主要材质是不同的平针织物与珠子，为了进行恰到好处的融合，我染了60多种颜色。我通过密集的卷边控制并延伸材料，制出卷、活褶与造型，代表神经细胞对人类运动及情感的控制。

I design for interesting women who are different from others, seek new challenges, appreciate life, desire freedom and emotions. My collection was inspired by a number of ideas: human life in the distant future, the development of brain neurons, and the fragility and strength of human emotions. The main materials of the collection are various types of jersey and beads, of which I dyed more than 60 colors to create the perfect combinations. I used intensive beading to control and stretch the material, creating volume, pleats and shapes in the clothing that represents the neurons' control over human movement and emotions.

灵感来源取材自水墨画的笔触,以大胆抽象的形式显示在衣服及物料上,寓意中国社会及经济发展一日千里,东西方文化及国际贸易交流让中国强大起来,同时传统观念产生蜕变,新与旧的价值观或在融洽,或在角力。有别于一般的时装设计在水墨画的演绎,设计师无论在印花上的构思还是物料上处理水墨的手法是非传统及破格的:设计构思、过程及作品是基于新与旧的价值观而成。

The inspiration comes from the brush strokes of ink and wash painting, showing it on the clothing and material in a bold and abstract way. It means that China is developing in a fast speed, and the communication between eastern and western cultures, as well as international trade makes China become stronger. Meanwhile, the traditional concept is changed, and new values and old values are in harmonious or against each other. Untraditional fashion design is illustrating the ink and wash painting. Fashion designers use untraditional skills in terms of conception of printing and ink painting handling of materials; the design concept, process and painting works are all derived from the new values.

香港，作为中国的一座重要城市，也是未来中国的一块瑰宝。以香港这座城市作为基点，发散至各个领域，并以其发展为出发点，对人与人，空间与人，以及人与社会阶层之间的相互关系进行重新思考，创作出一系列极具时代感的服装。

As an important city of modern China, Hong Kong is a treasure of the future China. In order to create a series of clothes with a strong sense of times, I rethink the relationship between human beings, space and human beings, as well as human beings and social classes, based on Hong Kong and spreaded to various aspects.

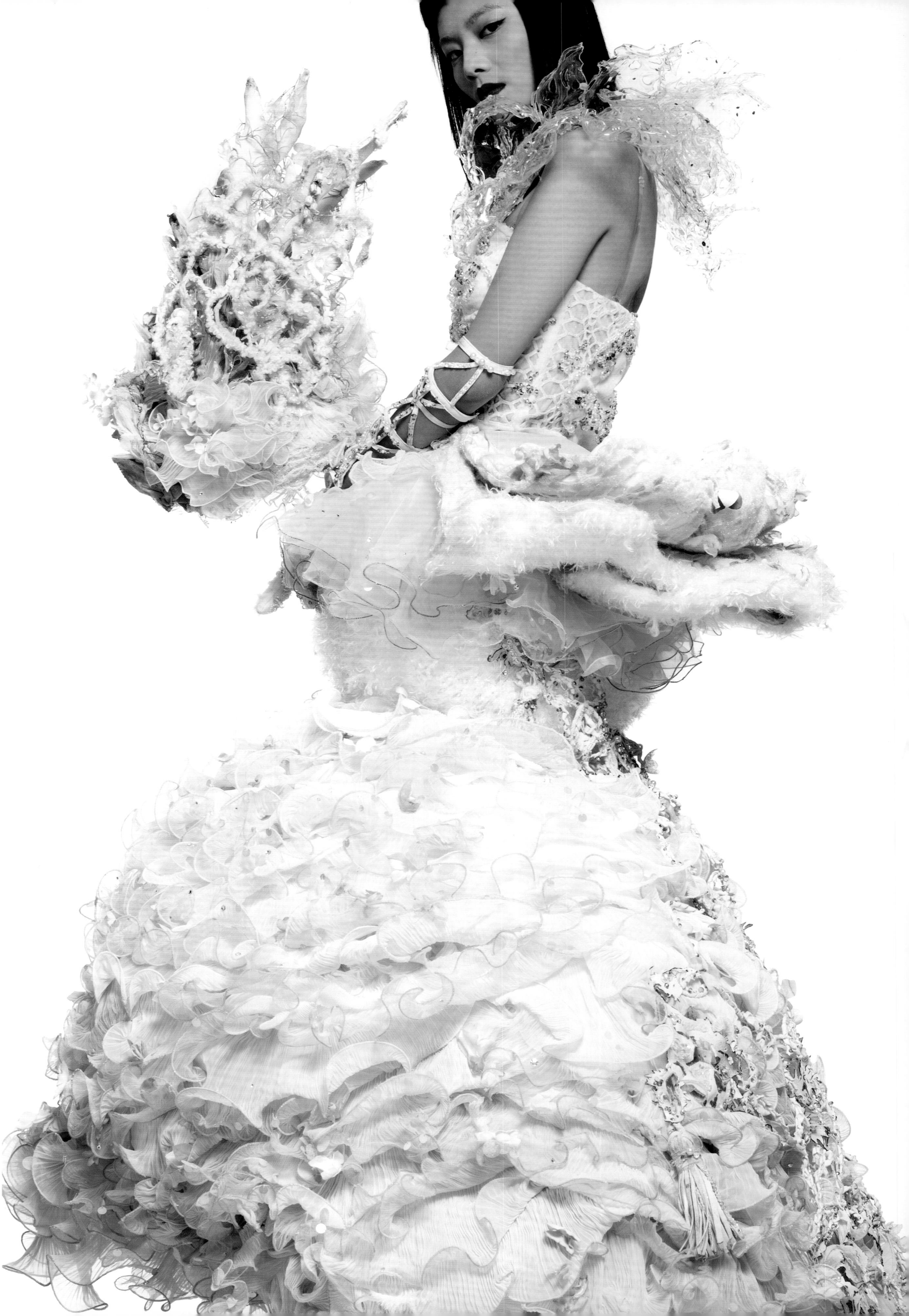

设计灵感来源于20世纪早期主要兴起于欧洲的未来主义艺术运动。未来派在运动以及运动轨迹中发现美。未来主义艺术运动时期的图片与绘画使用喷墨式打印技术,通过重叠透明材质呈现一种动态感。

This piece was inspired by the Futurist art movement, which occurred in the early 20th century, primarily in Europe. Futurism saw beauty in the dynamism of movement and paths of motion. Photographs and paintings from the Futurist art movement have been incorporated into the piece's textile using an inkjet printer, and a feeling of motion was brought to the design by overlapping transparent materials.

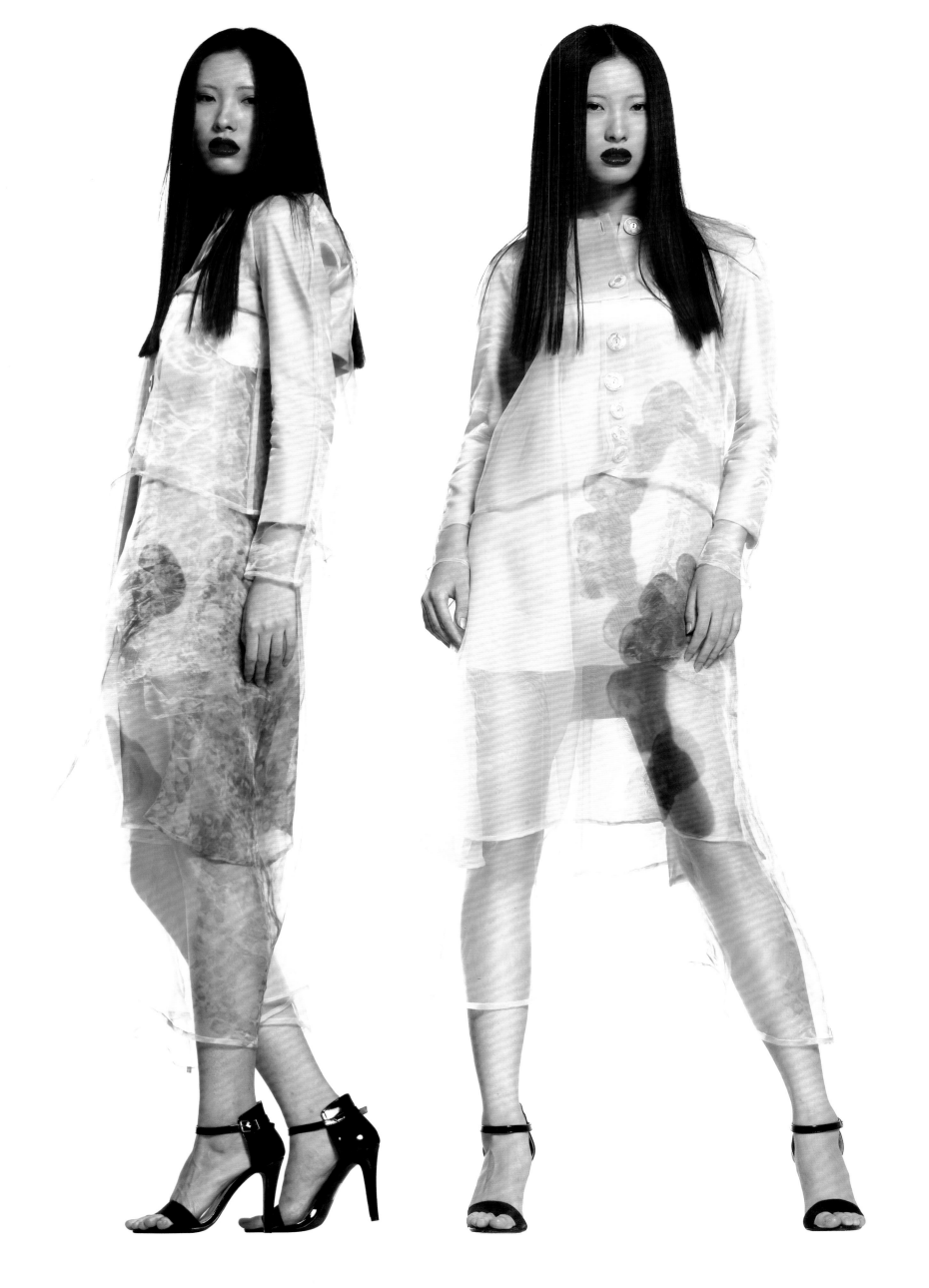

在不久的将来,环境遭受污染,在充盈着病毒和细菌侵害期出生的孩子,免疫系统将产生变异,以不断适应环境的改变。但是,也会有一些家长由于过分担心与猜忌而对孩子过度保护。这些家长不相信政府,也不相信制药业,将孩子置于无菌环境下,避免其与外界接触。因此,越来越多的孩子生活在安全地带,只能与家人沟通。但是,对他们的保护越多,体质会变得越弱。

本作品以两个女孩的生活情景向人们呈现与其他时光毫无差别的时光。

In the near future, at the age of polluted environment, there will be a need to develop protection against the harsh reality. Future children born at the times of artificial virus attacks and biological warfare will have mutated immune system constantly adopting to the changing conditions. However, there will always be overprotected parents full of exaggerated fears and suspicions. This generation of parents who won't trust neither the government, nor the pharmaceutical industry will put their children under the sterile conditions to avoid any contacts with the outside world. That is why from day to day more and more children will live in the secured homes and communicate only within their family circle. The more protected they are, the weaker their bodies grow. The project presents a day from two sisters' life, which is not different to any other days.

我在反复的循环里探索自己，同时在这一次次自我答辩的过程中去寻找出口的力量……"绣"本身就是一种宣泄与抒发，对于古代妇女而言，是以手代口地去"说"，而在此除了隐性的说，也同时在借由"绣"这般可见的图腾，一如外显的符咒般去召唤、去提醒自己去找寻自我……找寻自己认知好的衣服的可能。我把这整个过程视为仪式般的认真与入戏，同时借这场戏告知自己的蜕变。

I explore myself through repeated cycle, and summon up to search strength on the process of repeated self- defense…. Embroidery represents a form of unbosoming and expressing. For ancient women, embroidery means speaking using hands instead of mouth. Besides, apart from the recessive speaking, they are using visible totems like embroidery to summon and remind themselves to conduct self-quest, to find the possibility of wearing beautiful clothes…. I regard the whole process as a ritual and remind myself the disintegration through this play.

以仪式纪念那些介于男孩与男人之间的时光，并纪念那些我们如此斑驳却闪闪发光的岁月。以腰带铭记责任与负担，愿我们能永远记得此刻的自己。一瞬即永恒。

Commemorating the times between boys and men through ceremony, and commemorating those mottled but shinning days. Remembering responsibilities and burdens using waistbands, hoping that we can remember us at present forever. A moment is forever. Grow gold, then forever.

艺术设计作品
ART AND DESIGN

设计与艺术展区展示了来自10个国家和地区（其中包括4个APEC成员）的44位设计师和艺术家的雕塑、绘画、装置、陶瓷、首饰、服饰、创新面料、影像、动画、摄影作品和互动服装服饰产品。这些作品呈现了时尚语境下青年设计师对于生活方式的塑造和对于美的追求。

The Art & Design Static Exhibition displays the works of 44 designers from 10 countries and regions (containing 4 APEC members), including sculpture, painting, installation, ceramic art, Jewelry art, new textile, digital media art, animation and films, photography and interactive fashion design works. These young designers, with this products, illustrate their unyielding pursuit for the creation of the contemporary lifestyles and the inspiration for the ultimate beauty.

build a grid that works
with existing advertising
and remains flexible

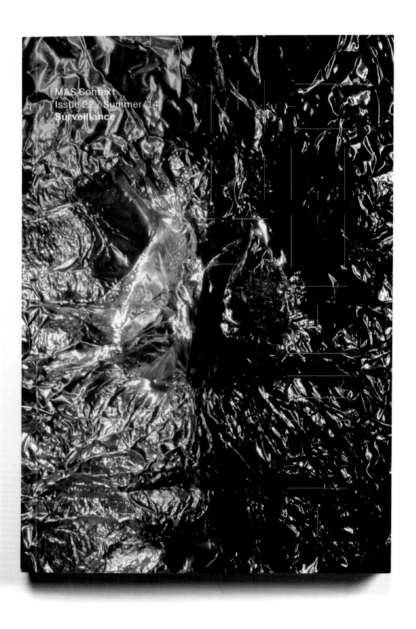

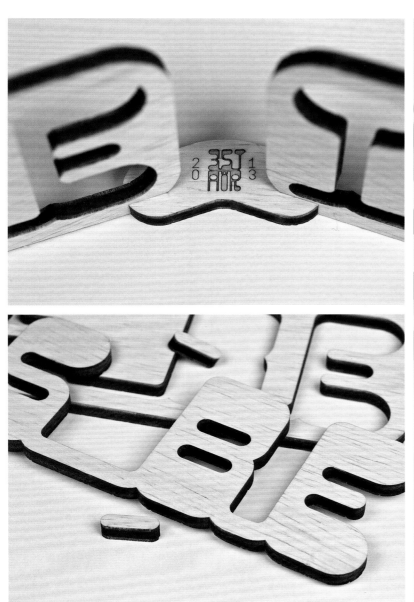

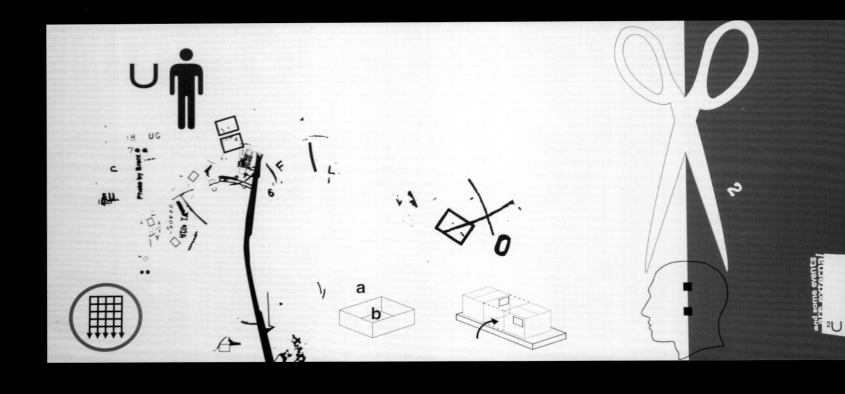

我的作品不断表达并代表我的想法，是我灵感的集合。首先是想象在某种程度上激发我将刺绣品焕然一新。我一直在为自己的作品寻找一种新的阐释意义。在技巧方面，把线作为媒介融合不同的时间空间。我并不担心会破坏一幅画，对我而言重要的是运用刺绣这种不经常在图画中使用的技巧，将其运用于图画中，因为这是过程的一部分。
通过对美的洞察，我将想象融入其中，有时是赋予作品一个新的身份或不同的美学概念。通过刺绣赋予想象一种新的情感、新的生活、新的美学阐释。

My work is a constant search to express and represent my ideas. My occurring artworks are a reaction of my inspiration. This starts with an image, that inspires me in certain way to do an embroidery that changes it into a new one. I am always searching for a new sense of interpretation for my pieces. My technique for that is using thread as the medium to merge different time spaces. I am not afraid of breaking a picture. The important thing for me is using embroidery, a technique not usually used on paper, to do it on a photograph, because this is a part of the process. I intervene images by applying my own perception of beauty to them., sometimes by giving them a new identity or a different aesthetic concept. It's the chance to give this image a new emotion, a new life, a new interpretation of beauty through embroidering.

贾大鹏是来自北京的潮流插画师和平面设计师，作品以高饱和度的色彩结合多变的场景，构成其无厘头的创作风格。作品曾发表在《亚洲青年设计平台2》，他曾服务于 Adidas Originals／Newbalance 等潮流品牌。

Jia Dapeng is a fashion illustrator and graphic designer coming from Beijing. His work combines high SAT colors and poly tropic scenes to create the reasonless style. His work once published in Asian Youth Design Platform 2, and he served fashion brands such as Adidas Originals Newbalance.

此组作品描绘的是 Prince、Machael Jackson、David Bowie 三位音乐家，这三位艺术家是华丽摇滚的代表人，也是流行文化最重要的先锋。流行文化的概念几乎涵盖了公共空间的所有事物，它是真实生活的写照，并且用最华丽的形式表现了相似的现实。它遍布我们周围，影响我们的思维、感受以及无数的生活方式。通过这组插画作品诠释我对流行文化的理解。

The work portrays three musicians-Prince ,Machael Jackson, David Bowie who are the representatives of Glam Rock as well as the most important pioneers of pop culture. The conception of pop culture almost covers all the things in the public space, and it is the portrayal of true life and expresses similar reality using the most gorgeous style. It is widespread around us, and it influences our thoughts, feelings and life styles. I want to illustrate my understanding toward pop culture through this work.

通过探索自然界，比如重复与变化规律，将这些体验与母体文化研究和旅美、旅澳的生活经历相糅合；在这三者之间寻求设计理念和创作方法，使得金属工艺与艺术首饰创作富有个人风格，体现了人文与自然的多重特性。

The experience, such as the rule of repetition and transformation, combined with native culture investigation and travel experience of America and Australia inspire the design concept and creation method. This makes the metalwork and artistic jewelry design full of personal style which reflects the macro feature of culture and nature.

可穿戴设计目前主要是手环和眼镜,背包作为一个穿戴产品并且有实际储物功能,也能智能化去设计。我从运动人群的特点和生活习惯上去寻找设计点,发现这个人群在一些背双肩包的特定场景下,有一些诉求是可以通过把背包智能化来满足的。面对传统的背包加工企业,他们也需要一种新的产品、新的使用方式、新的交互方式、新的商业模式,智能背包是一个方向。

Currently, wearable design mainly focuses on bracelets and glasses. As a kind of wearable product with actual storage capacity, backpack also can be designed intelligently. I search for opportunities from the characteristics and living habits of those who like sports, and I find that some requirements of those people can be satisfied by the intellectualization of backpack. Faced with traditional knapsack processing enterprises, they also need new products, usage modes, interactive modes and business models. Intelligent backpacks are the direction.

其灵感来源于中国的牡丹花，象征着富贵吉祥与享乐。结合了传统錾花手工艺与当代工业电铸技术，以及传统题材与当代结构设计，此系列源于古典文化却带有当代气质。在尝试新的珠宝制作方式的过程中，完成外形夸张却佩戴轻便的项饰。

The inspiration comes from peony, which represents wealth, propitious and enjoyment. The work combines traditional carve handicraft and modern industrial electroforming technology, and combines traditional subjects and modern structural design. This series is originated from classical culture, but with modern disposition. While trying new jewelry production method, I complete the crests which are exaggerated in appearance but lightweight to wear.

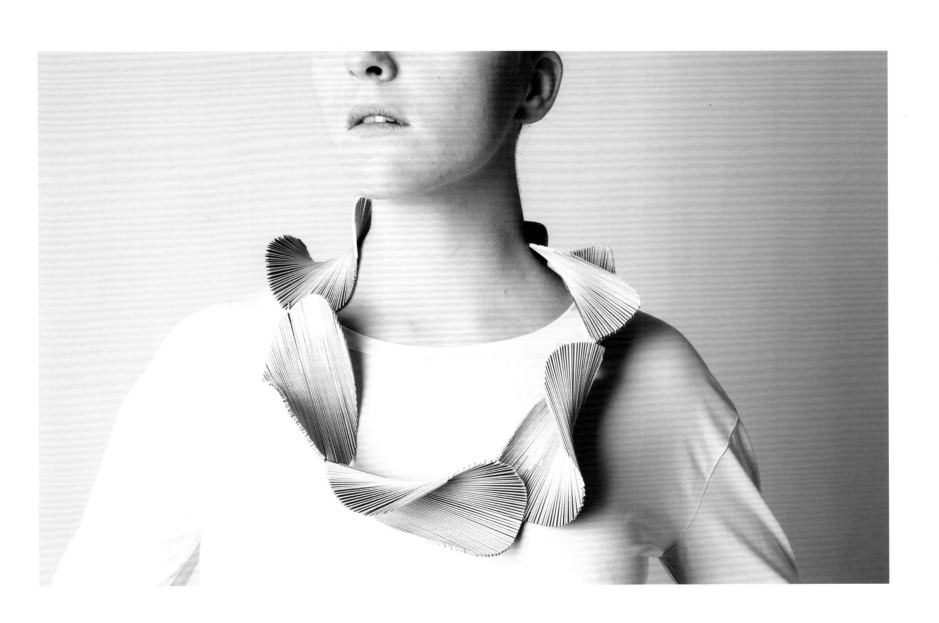

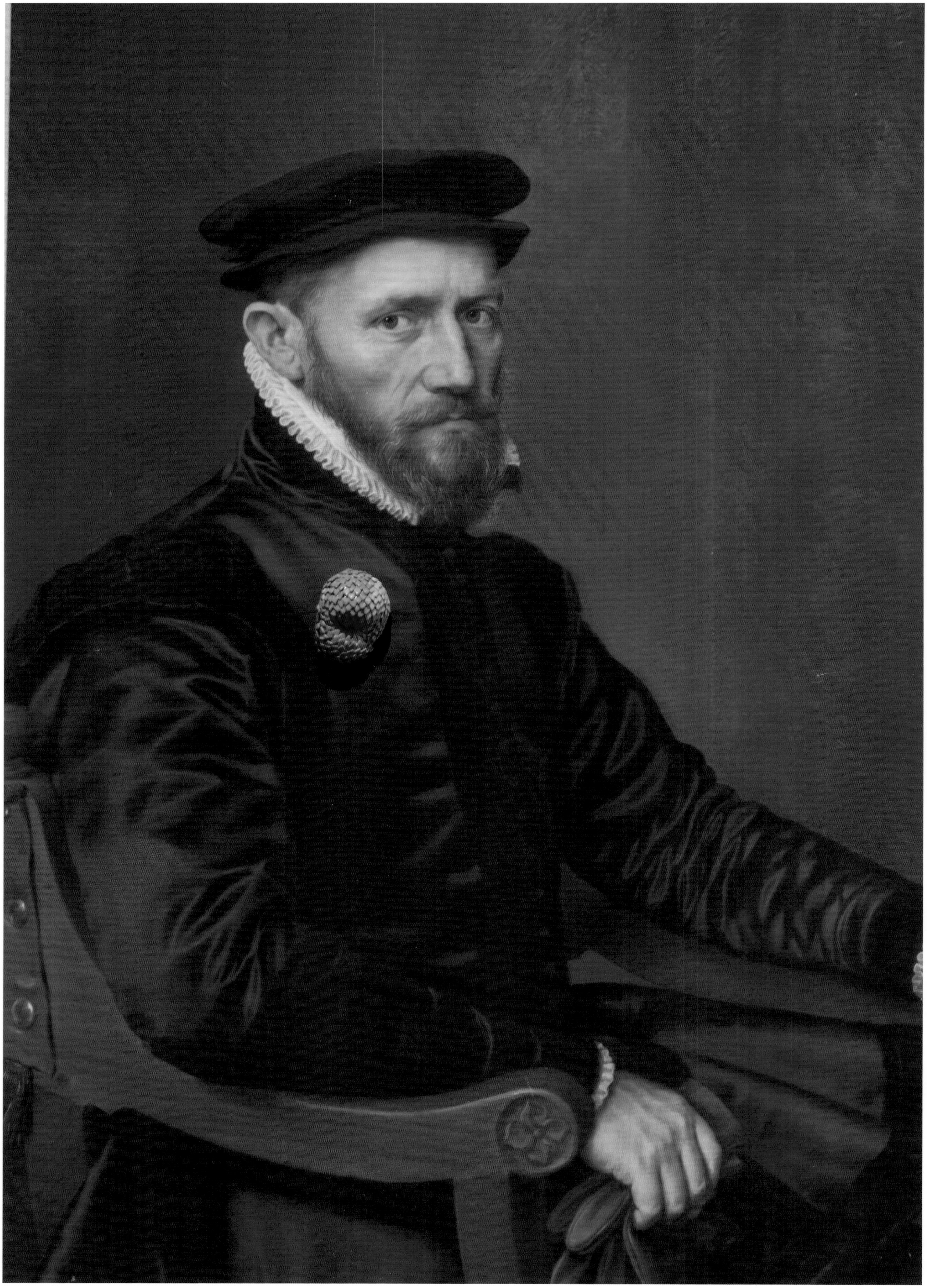

中国有句俗语："面由心生"。人的面部表情是通往其内心深处的钥匙，对面部的描述是一件神秘又富有挑战性的事情。在艺术创作领域，对面孔的描述大多呈现于绘画、雕塑、摄影等方面。中国古代皇帝都会将自己的肖像画流传于后世。古罗马时期，面容雕塑是主流，如罗马皇帝君士坦丁一世和狄奥多西一世的肖像闻名于世。对"面孔"的表现成为艺术家自身探索和表达自我内心世界的方式。这个系列的胸针作品正是在对面孔抽象后，对当代首饰设计语言的诠释。

An old saying goes that "State outside is based on mind inside". Human beings, facial expression is the key to the cockles of the heart. Therefore describing the facial expression is mysterious and challenging. In the realm of artistic creation, the description of faces is mainly reflected in paintings, sculptures and photography works. In ancient China, the emperors would pass their portraitures on to their later generations. In ancient Rome, face sculpture took the domination. For example, the portraitures of Roman emperors-Constantine the Great and Theodosius the Great were world famous. Expressing "faces" becomes the method to explore and present the inner world of artists. This series of brooch work is my way to interpret modern jewelry design language after abstracting the faces.

德国新天鹅堡是世界上最著名的童话城堡，它是迪士尼标志的原型，以及睡美人城堡的原型。2013 年的圣诞节，叶子乐带着由英国戏服设计师特别定制的戏服和他的团队来到新天鹅堡，在冰天雪地里拍摄了这组睡美人的故事。

Germany's New Swan Castle is the most famous fairy tale castle.It is the prototype of Disney symbol as well as the Sleeping Beauty Castle. On the Christmas of 2013, Ye Zile and his team went to New Swan Castle with the special costume designed by British designer.They shot the series of Sleeping Beauty story on the frozen and snow-covered land.

德国新天鹅堡是世界上最著名的童话城堡，它是迪士尼标志的原型，以及睡美人城堡的原型。2013 年的圣诞节，叶子乐带着由英国戏服设计师特别定制的戏服和他的团队来到新天鹅堡，在冰天雪地里拍摄了这组睡美人的故事。

Germany's New Swan Castle is the most famous fairy tale castle.It is the prototype of Disney symbol as well as the Sleeping Beauty Castle. On the Christmas of 2013, Ye Zile and his team went to New Swan Castle with the special costume designed by British designer.They shot the series of Sleeping Beauty story on the frozen and snow-covered land.

这部短片是受法国品牌 Each x Other 委托制作，Each x Other 受周依的 3D 动画作品启发制作了一个服饰系列。当周依第一次想到 Each Other 这个词时，马上就联想到很多人、事、物、地点以及不同情感层面以时间错落的顺序融合在一起。就像跟着梦境中的错落片段一样，随机性和潜意识都融合成一体。就像一个自由风格的音乐剧或即兴舞蹈，在"Hollowness"中，这些陌生人、真实抑或虚构的人物都来到一起。这也被周依称为新的民主，只能在社会媒体舞台上找到的民主。

Steve Mcqueen 佩戴的意大利标志性眼镜品牌，同时也是周依长期热衷的品牌 Persol。邀请周依编写一则故事并以艺术家的身份独立拍摄短片之时，与王子、王妃的相遇激发了周依无限的灵感，她写出了可以由 Clotilde 与 Emanuele 演绎的全新社会媒体故事。这也将是 Clotilde 与 Emanuele 首次共同出演的影片。他们诠释了一个具邻家男孩女孩特质的现代童话，将神秘的王子与公主身份带到我们日常生活中。

2013 年 5 月 5 日至 6 月 5 日，《文化香奈儿 N°5》(N°5 CULTURE CHANEL) 展览于位于巴黎的东京宫当代艺术中心（Palais de Tokyo in Paris）展出。多媒体艺术家周依受香奈儿 5 号之邀请，携带自己为此次展览制作的影像《周依眼中的 CHANEL N°5》出席。

The short film is entrusted by the French brand Each x Other, which was inspired by Yi Zhou's 3D cartoon work and created a collection. The first time when Yi Zhou thought of the word "Each Other", he immediately mixed many people, many things, places and feelings at different levels in chronological order. Just as the scattered fragments in a dream, randomness and subconscious are mixed together. Those strangers, real or imaginary persons come together in the "Hollowness" like a free-styled music drama or impromptu dance. Yi Zhou also calls it the new democracy, which can only be found in social media stage.

The Italy symbolic glass brand Persol which is worn by Steve Mcqueen, is also the brand that Yi Zhou shows zeal. While inviting Yi Zhou to write a story and shoot a short film independently as an artist, she was inspired by the encounter of prince and princess. She wrote a new social media story which could be displayed by Clotilde and Emanuele. And this will be the first that Clotilde and Emanuele play the same film. They illustrate a modern tale with the characteristic of boys and girls next door, bringing the mysterious status of prince and princess to our daily life.

N°5 CULTURE CHANEL exhibition was exhibited in Palais de Tokyo in Paris from May 5th, 2013 to June 5th, 2013. Yi Zhou, the multimedia artist was invited by CHANEL N°5, and attended the exhibition with CHANEL N°5 seen by Yi Zhou, which was made by him specially for this exhibition.

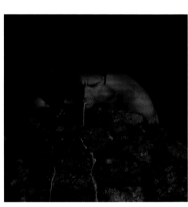

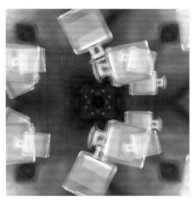

按照我个人的审美情趣,我经常会被以图形和文字为主的海报所吸引。根据这种独特的风格,我以"未来即现在"为主题,创作了三款海报系列。在该系列中,"下一个"是用最小的图标对排版进行探索,进而阐释三类发人深省的观点,即未来将会是什么样子的。该作品意在使观者产生反应,吸引他们思考每个观点,想象自己在这些观点的影响之下会得到何种结果。

I have always been drawn to graphic and text-focused posters when it comes to my personal aesthetic. Following this distinct style, I have created a series of three posters that go along with the topic of "Future of Now". This series entitled "Next" is an exploration of typography with minimal icon use to help illustrate three thought-provoking statements that centre around the notion of what lies ahead, what our future holds. This piece is meant to obtain a reaction from its viewer, and tempt them to consider each statement differently, allowing them to imagine where each message can lead them to.

灵感来源于竹工艺以及东方文化，通过改变材料的原本形态实现设计。我所创造的材料用于制作当代配饰。

Chinese and Japanese bamboo craftsmanship and oriental cultures inspire me to manipulate materials and change the normal state of materials. The materials that I created were then translated into a collection of contemporary accessary designs.

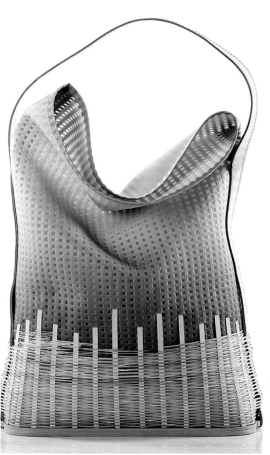

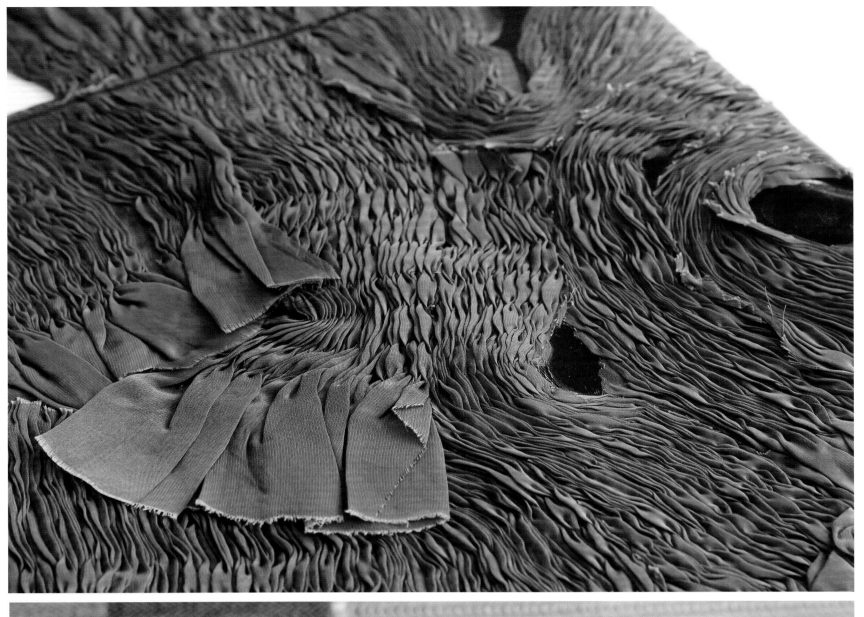

首饰艺术来源于生活,艺术拥有了生命,可以展现生活中生命的意义。生命的象征性特征之一是"动",所以在整个系列中用"动"的形式作为主要的结构特点。

这个系列的主题为《"SPACES"&"STEPS"》(空间和媒介),SPACES 的意义:有了空间,事物才能够存在。人生存于空间里,人的一切活动都在空间这一个"容器"当中。人与空间的关系是"沟通"的关系。STEPS 的意义:人与空间的沟通是通过媒介实现的,所有作为人与空间的媒介都是"STEPS",人本身的活动能力,就是一种与空间沟通的媒介,不过人的身体本身无法满足人与空间沟通的需要。生活中几乎所有事物都是人与空间沟通的媒介——"STEPS"。"STEPS"和"SPACES"构成了人的生活。在整个设计系列中,用艺术化的形式展现了"PEOPLE""SPACES"与"STEPS"的关系。

Jewelry art is derived from life. Art can reveal the meaning of life once it gains life. One of the symbolic features of life is "act", therefore I use the form of "act" as the main structural feature through the whole series. The theme of the series is "SPACES"&"STEPS". SPACES means that things can exist only if there's space. People live in spaces and they act in the spaces. The relationship between human beings and spaces is "communication". STEPS means that human beings and spaces communicate through "mediums", and all the mediums are "STEPS". The mobility of human beings is a kind of medium that communicates with the spaces. However, human bodies can not meet the need of communication between human beings and spaces. Almost all the things in life can be the communication medium between human beings and spaces-"STEPS". "STEPS"and "SPACES" constitute people's life. I demonstrate the relationship between "PEOPLE", "spaces" and "STEPS" through the design series.

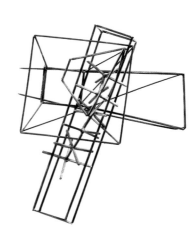 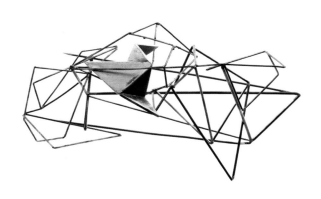 

在每个珠宝饰物中，我可以创造一个世界，也能够毁掉一个世界，可以按照自己的想象畅说最不可思议的地方。我喜欢艺术家是说故事的人这一角色，是创造者又是编造者。能够让观者相信这些传说便是最大的本领。首先吸引他们的眼睛，其次是心灵。

珠宝饰物是想象力的胜利品；是两类人的世界——艺术家与观者，这两类人能够瞬间融为一体。

In a piece of jewellery, I can create and destroy world, talk about the most unbelievable places to which I have drifted in my thoughts, and talk about who I have seen there. I like the role of the artist as an ancient storyteller, the creator as liar. Their ability to make the viewer believe these tales is the greatest talent. To first take their eyes and, by that means, their heart.

Jewellery is a triumph of the imagination; the world of two people -the artist and the spectator-they may integrate each other in an instant.

 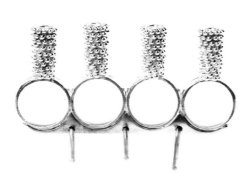 

China 中国 | Tang Tian 唐天

两件作品均出自我的 Petty（小宠物）系列。每件作品都代表一个小宠物，分别象征了我生命中遇到过的几位真实的女性形象。作品灵感源于我认为每一位女性都是一个琢磨不透的奇特生命体。女性要懂得保护自己的独特和可爱。

The two works belong to my petty Series. And each work stands for a petty, representing several true female images I have met in my life. The inspiration comes from the fact that I believe every female is a unique life entity which can not be understood. Females should learn to protect their unique and lovely characteristics.

China 中国 | Hu Naishu 呼乃纾

该设计研究了融合纱线及其编织过程中结构与外形的特性、反应以及影响。整个过程重点在于编制结构的发展变化，合成材料或天然材料的组成，加工工序，加热时复杂的面料操作。设计灵感来源如下：所选的材料、通过编织设计创造一种新的表达方式，使面料的金属坚硬外形与柔软面料之间形成对比。

This project investigates the study of melting yarns, and their properties, behavior and impact on the structure and appearance of the weaving. The process focuses on the development of the woven structure, the composition of synthetic and natural materials, the finishing processes, and complex manipulations of the fibers through heat. The project was inspired by several components: the chosen material, the intent of creating a new language of expression in woven design, and the unique dichotomy between the look of the metallic, rigid surface versus the soft feel of the textile.

我设计的珠宝饰物体现于互相吸引与排斥之间的张力。使用不相称的材料,将普通物体变得不同寻常。在18世纪,发现了许多新品种的动、植物,而且那时人们好奇心极其强烈。人们搜集珍品,因为能够给他们带来一种特别的感觉。虽然没有发现新物种,但是我可以通过使用生物的尸体唤起那种奇妙的感觉。使生物得以重生,被赋予令人瞠目的新鲜生命。

My jewellery deals with the tension that lies between attraction and repulsion. I take inappropriate materials, making ordinary and familiar objects seem extraordinary and unfamiliar. In the 18th century many new breeds of animals and plants were discovered and it was the main era of cabinets of curiosities. People collected rarities because it gave them the feeling of being in the presence of something extraordinary, In a world, where not many new breeds are discovered, I use dead creatures in my pieces to evoke wonder. The creatures are transformed and reborn; and given a new life as objects of astonishment.

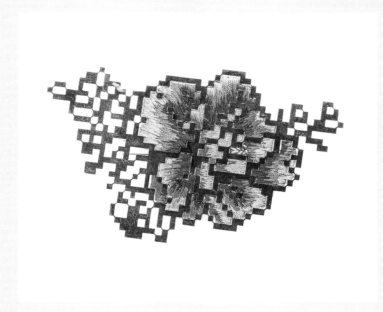

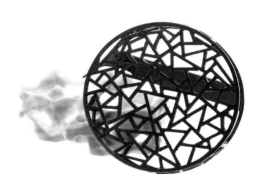

造型艺术作品
PLASTIC ART DESIGN

该组作品是我 2014 年的毕业设计作品，以建材金属、废旧金属、机器零件等为材料，通过焊接的构成方式独立完成。该系列作品由五只蜂和一个蜂巢组成，以昆虫（蜂）的形象进行具象创作，以废旧金属极具表现力的独特性丰富了该作品的视觉呈现，从侧面隐喻了工业文明与生态文明的矛盾关系。之所以取名"杀人蜂"，它的深刻含义是工业文明作用下的生态环境日益侵蚀及危害人类本身，希望借"杀人蜂"之名引起观者内心的反思。

This is my graduation design work of 2014. The materials include building metals, scrap metal and machine components. And I completed the work by way of welding independently. This series consist of five bees and one honeycomb. I carried out concreted creation based on the image of insects (bees), and the uniqueness of scrap metals enriched the visual effect of this work, metaphorizing the contradictory relationship between industrial civilization and ecological civilization. I name it as "Killer Bee", because I want to cause human beings reflection under the background that the ecological environment is harming mankind due to the development of industrial civilization.

作品《竞争与遗迹》展现食物之间相互竞争水分的过程与结果。食物枯萎、腐烂凋败，留下的是一片灰白色的、带有食物表皮肌理质感的容器胚子。作品试图发问地球村的居民如何在此星球上竞争或共存？作品获得2013年度core77食物艺术与设计奖。

The work "Competition and Relics" shows the process and result of competing water among foods. The result is that the foods withered and decayed, and only the vessel cotyledons in grey white with the texture of food skin remained. This work is designed to indicate that how should human beings compete or coexist in this planet? The work "Competition and Relics" won the Core77 Food Art and Design Award in 2013.

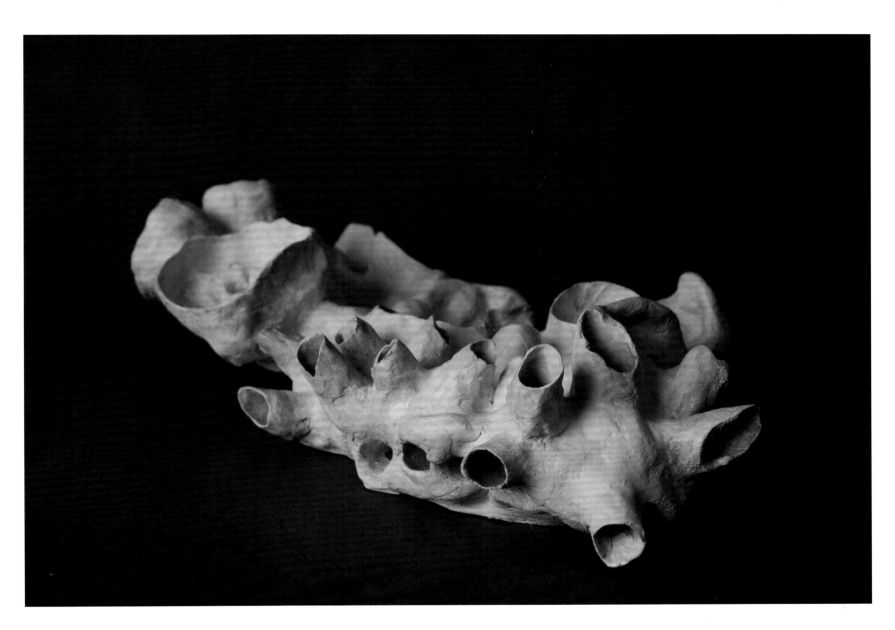

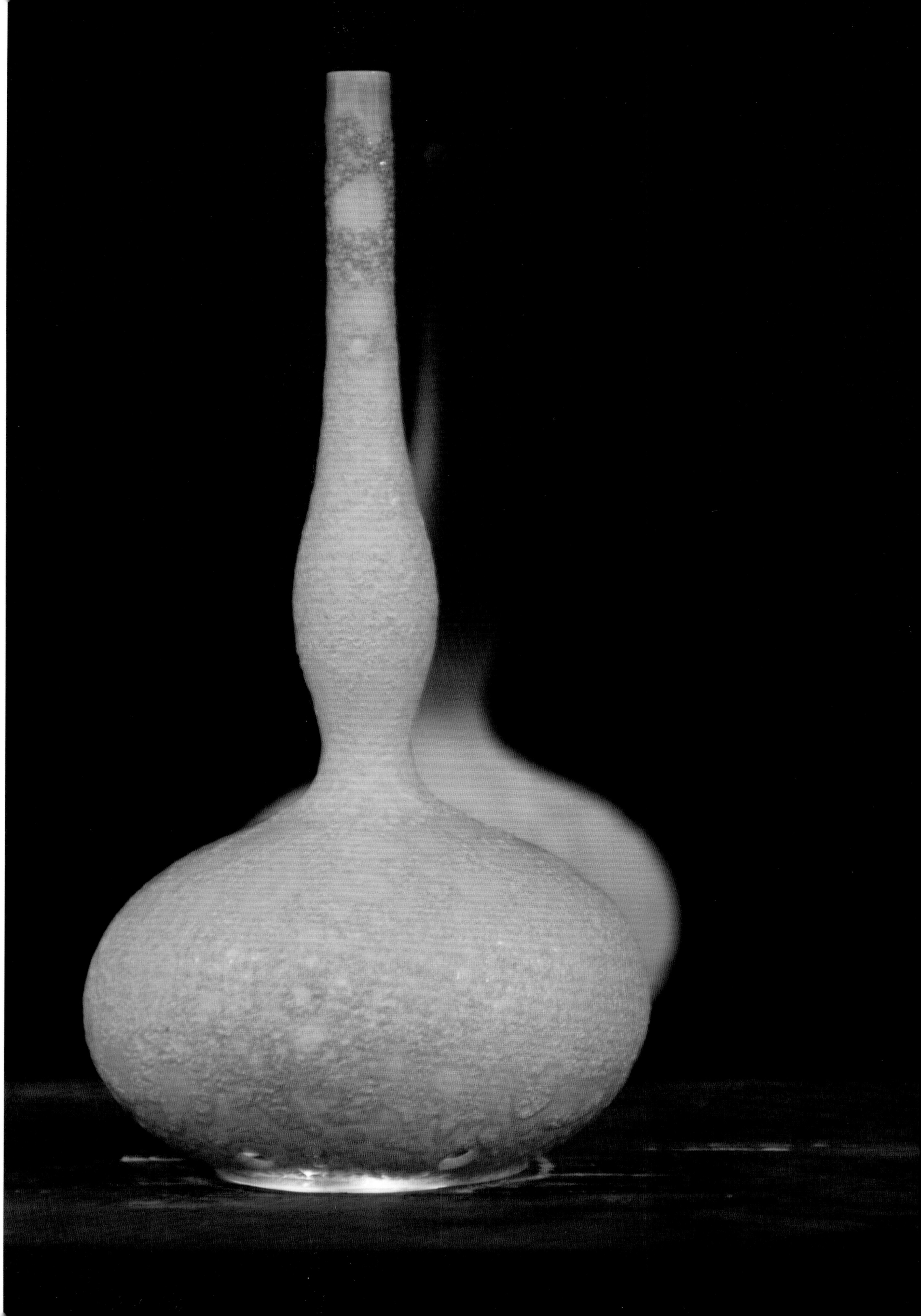

霓裳系列作品以现代时尚的生活为主题，对女性人体的概括、变形，对衣纹的独特艺术处理，对材质的着色、磨光，让材料说话，使雕塑达到难于用语言表达的意境。时尚是什么？在材料的运用、线条的表现和形体的塑造中已经表达了作者对时尚的精神本质的诠释，更给时尚一个全新的感觉。同时，也表现了当下璀璨旖旎的时尚生活下人们的生存状态。

The theme of the Series of Seduction is modern fashion life, through the generalization and transformation of women's bodies, the unique artistic processing of clothing texture, as well as the coloring and polishing of materials. This sculpture achieves an ideal condition which can not be expressed by words. What does "fashion" mean? The designer has already expressed his understanding of "fashion" through the usage of materials, the expression of lines and the shaping of bodies. Besides, the designer endows "fashion" a kind of new feeling, and indicates human beings' living condition under the bright and charming "fashion" style.

《罔山》灵感来源于中国山水画,把山水画中的文人意境延伸到雕塑形态中,把平面的山脉转化成了有层次的立体造型。在形式上,吸取"文人园林"的苏州园林"窗格"这一元素,在曲折多变与虚实相间,窗内与窗外相互渗透中,借景精心剪裁出一幅幅充满人文写意的山水画。作品用瓷作为材料,能更好地传达出其中"罔"的意境和内涵,营造出一种冷静孤寂与单纯脱俗的气质,也是对当下这个时代尘俗、烟火、急躁的社会状态的一种反思。

The work is inspired by Chinese Landscape Painting. The designer extended the literary artistic conception of landscape painting into sculpture forms, converted flat mountains into stratified three-dimensional shape. In form, the designer used the element of "pane" from the "literary garden"- "Suzhou garden". Among the winding changes and false or real conditions, the inter infiltration inside or outside the windows, the designer created landscape paintings filled with literary elements. This work used "porcelain" as material in order to better convey the artistic conception and connotation of "Wang" and create the solitude and refined temperaments, which is also a reflection of the current worldly and impetuous social status.

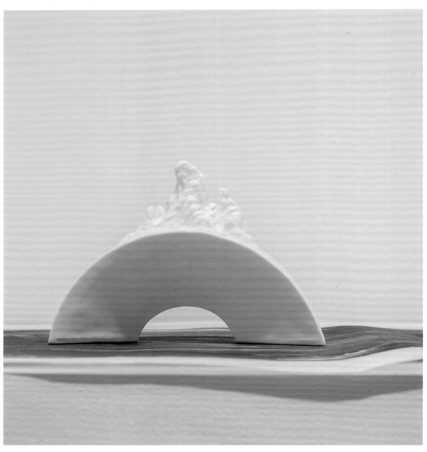
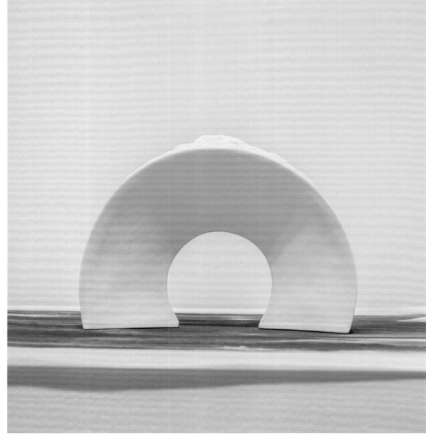

陶瓷作品《家》，是我自身生活的缩影，是我对家的向往。这里有我童年的影子，也有未来对于家的畅想。作品运用写实手法，高温的釉色变化赋予了作品生活的痕迹，像发了黄的黑白老照片，斑驳的效果仿佛将时间凝固在了里面。

The ceramic work "Home" is the epitome of my own life and my desire toward home. There's my childhood's shadow and imagination of the future home. The work adopts realistic approach. The work looks like black and white pictures turning yellow, and is endowed with life traces under the hyper thermal glazing color changes. It seems that the mottled effects reflected through the work solidified time.

创意说
IDEAS

Kim Shui
美国
America

我开始在法国学习经济学，但是我对时尚非常有兴趣，所以我想追求更多的方面，于是我选择了两个交流背景，商业和时尚。

我也会使用一些混合的自然材料，比如动物的皮毛和亚麻。我也会运用塑料，因为塑料具有历史性和特殊的协调性。

我认为这是一次非常棒的经历，我很荣幸来到北京，并且来到这所优秀的学校——北京服装学院。在这里，很好的体验就是，这是一所很棒的学校，这里的学生也很出色。看到北京服装学院这些有梦想的学生，我受到了极大的鼓舞，所以这真的是一次很棒的经历。

At the beginning, I studied economics in France. But I'm interested in fashion, and I want to pursue more aspects, so I chose both business and fashion.

I also use some mixed natural materials, such as animals' wool and flax. I also use plastics, which is historic and coordinated.

I think the experience is excellent. I am honored to be invited to Beijing, especially to this outstanding school-BICT. What impressed me most is that the school is excellent and the students are outstanding. I am greatly inspired by those students who have dreams. So, it is a good experience.

克里斯·拉恩·林
Chris Ran Lin
澳大利亚
Australia

Hello，大家好，我是来自澳大利亚的华裔设计师Chris，这次很高兴能收到北京服装学院的邀请来这里展出我的最新作品——软雕塑。

作品灵感是通过对不同材质面料的破坏和重新雕塑，去表达我对男装里面阳刚和柔美的不同表现。这次作品运用羊毛质地的材料，因为我觉得羊毛材料本身已经充满了力量感和很强大的可塑性。作品里很多是以针织为主，我觉得针织的质感给人传递了一种很柔美很实在的感觉，更重要的是也传递出一个信息，那就是我们可以通过一些像布料面料这类软性的材料，做出可穿性的雕塑作品。

我觉得每个设计师都有自己的特点，因为设计师的作品如果有自己鲜明的风格，能够让人一眼就认出是谁的作品，那是非常成功的。这次我看到的设计师的作品都很不一样，我是主要做男装设计的，但是看到一些女装在工艺、细节上都不错，很有自己的想法。

这次参展也是一个很好的机会把我的作品呈现在中国观众的面前，同时也看到了其他地区设计师的作品，我觉得是一个很好的交流方式。其实也很感谢主办方给设计师安排的助理团队，通过他们我能更好地了解北京当地的一些文化和很地道的东西。希望以后能有更多的机会带着我的作品来到中国、来到北京，和大家见面，谢谢！

Hello, I'm Chris, a Chinese designer from Australia. I feel it a great honor to be invited by BIFT to exhibit my new work-Soft Sculpture.

There is via of folding, breaking the fabric to express the manly and graceful of men's wear. I use wool materials in my works, because I think this kind of material is filled with strength and plasticity. Many works are fixed on knitting, because in my point of view, the texture of knit enables people to feel graceful and real. Besides, it tells us that we can create wearable sculptures using soft materials like cloth or fabric.

I think every designer has his or her characteristic. If the work has its own distinctive style, and we can figure out the designer after seeing it, then the work must be a success. I find that all the works have their characteristics. Although I focus on men's wear, I also find that a lot of women's dresses are quite excellent in craft and detail with many ideas.

I think this exhibition is also a chance to show my works to Chinese people. Besides, I saw many works from other countries, which is a good way to communicate. I want to express my thankfulness to the sponsor for arranging assistant team to the designers, this team told me a lot about local culture of Beijing. I hope that I can have more chances to come to Beijing with my works and meet all of you. Thank you!

弗拉基米拉 B. 斯特福克
Vladimira B. Steffek
加拿大
Canada

事实上，我在中国碰巧买了一套豪华婚纱，在西方国家，有一种特殊的习俗：让人们忘记婚姻生活或者年轻时的自由快活的生活都是不可能的。第二次婚姻与第一次存在着很大不同，当你的新郎为你寻找第二次或者第三次的婚纱时，想要找到一件与第一次婚礼一样正式的婚纱是相当难的。

设计师的智力、个人经验以及个性影响着他设计历程。除此之外，他（或她）的生活方式和人生背景也会影响其设计，因此这意味着每逢设计师心情愉悦时，都可能诞生出与众不同的优秀舞台作品。设计师在其职业生涯中都会经历一段困难期，因此设计师的作品深受其生活的影响。我觉得另一个问题就是，在什么情况下你应该学习所看到的东西。也许今后生活可能会发生变动，但我会依然专注于我的设计。

Actually, I bought a luxury wedding dress in China by chance. In western countries, there's a special custom: it is impossible to make them forget marriage or the free life as a young man. The second marriage is quite different from the first one. While your bridegroom finding the second or third wedding dress, it is very difficult to find one that is the same formal as the first one.

His design process is influenced by his intelligence, experience and personality. Besides, the design is also influenced by his or her life style and background. Therefore, when the designer is in good mood, he is more likely to create distinctive stage works. All designers can experience difficult period, therefore they are deeply influenced by their lives. But I think the key point is what can you learn form those difficulties. I will focus on my design regardless of any changes in the future.

白鸽
Bai Ge
中国
China

HELLO，大家好，我是设计师白鸽。我是2011届北京服装学院的毕业生，毕业之后的几年里我去了伦敦，并就读于英国皇家艺术学院攻读女装设计的研究生课程。

我的设计灵感有很多，其中最主要的是我个人比较偏爱的一些有关自然和传统的东西。到了伦敦之后，不同的文化氛围也对我产生了不同的影响，使我个人的设计风格也有了不小的改变。我逐渐开始注意一些西方的艺术家，特别是当代艺术家的作品，他们给了我很多的灵感，所以在我设计的最新系列里，我想尝试把传统元素，尤其是中国的传统服饰文化元素融合到当代的艺术风格和艺术审美中。所以这次算是以我的毕业设计作为我的第一次尝试。

这次参展的是我在皇家艺术学院的毕业设计，是为2015年春夏系列设计的，我想把一些很传统的元素提取出来，包括中国的传统审美，把它们变得更加现代化，让它们可以融入到现代的生活当中去，而不是仅仅重复传统的一些元素或者只是把传统元素强硬地加到现代设计作品当中。大概流程是建模、打印、抛光、上色。眼镜上有许多小洞，佩戴者可以透过这些小洞来观察外面，以此来呼应古时的花窗。

Hello, my name is Bai Ge. And I graduated from Beijing Institute of Clothing Technology in 2011. After graduation, I went to London, and studied in Royal College of Art for postgraduate program, majoring in women's wear design.

I have many design inspirations, among which the most important is that I love things which related to nature and tradition. After arriving at London, I was influenced by different cultural atmosphere, thus my design style also changed greatly. I began to pay attention to western artists, especially works of modern artists. They brought me a lot of inspirations, therefore I try to combine traditional elements especially Chinese traditional fashion culture elements with modern artistic style and artistic aesthetics in my newly design clothes. So, I regard my graduation design as my first attempt.

The work is my graduation design of royal College of Art, and it is designed for the 2015 Spring/Summer series. I want to extract some traditional elements, including Chinese traditional aesthetic elements, to modernize them and combine them with modern life, instead of repeating traditional elements or put them into modern design works constrainedly. The process includes modeling, printing, polishing and coloring. There are many holds on the glass, so people can observe the outside while wearing it, and it is designed to echo the lattice window in ancient times.

狄梦洁 / Di Mengjie
中国 / China

很高兴和大家在北京相聚，希望大家都有一个很愉快的时光。我以前是 Stylesight 的男装部插画师，负责整年的流行预测。在 Stylesight 和 Wgsn 合并之后，我的职务是 wgsn 的男装插画师。

很多时候，我的设计灵感来自于我画的画，就是我有一种想法用画来表达，然后一看这个感觉很好，就做成衣服了。设计有很多不同的表达方法，有些人做设计，他对设计灵感的收集时间是很长的，他有灵感，有一个大的方向，然后在大的方向上找一些图片，用图片的方式表达一个故事，再从故事传达一些设计的细节，最后从细节传达设计的一个系列。有些人做设计是从模特直接入手，很随意，用布的质感去体现你想要表达的造型。

I'm glad to meet you here, and I hope that all of you will enjoy a good time. I used to be an illustrator in Men's Wear Department of Stylesight, and my main work was to predict the fashion elements of the year. After the combination of Stylesight and Wgsn, I became the illustrator in Men's Wear Department of wgsn.

I am usually inspired by what I draw, that is to say if I have an idea I will express it by drawing. If I think it feels good, then I will design it into clothes. There are many methods to express design. Some designers will spend a lot time selecting design inspirations. If he/she has an inspiration, or a wide scope direction, then he/she will find pictures and express a story by means of pictures. After that, he/she will convey some details through pictures, and then convey a collection through details. Some designers use models directly, and I think it's very random. They use texture to indicate the sculpture they want to express.

孙馨 / Sun Xin
中国 / China

这个系列的主要灵感来源于我对中国 20 世纪 50～60 年代时期的服饰研究。在我的作品当中有一个比较明显的细节，就是衬衫领子。我们可以看到中国 20 世纪 50～60 年代的历史文献，那时工人和农民都会拿一条毛巾在脖子上打一个结，这就是我的那些衬衫领子的设计来源。我希望人们穿上我的衣服会发现一些有趣的细节。

其实，很多时候灵感并不是从寻找当中得到的，很多是从生活中意外获得的，我觉得工作方法没有一个固定的模式。我会随时保持对周围发生的事情的敏感度，然后把它们运用到我的设计当中来。如果很多事情都是听别人说的，自己没有尝试过，对自己的人生就是一种遗憾。年龄对于设计师来说是一笔财富，很多艺术家，年轻的时候和比较成熟的时候的作品有非常大的区别。随着年龄增长，看事情的角度和成熟度，包括价值观念，都会发生变化。设计作品的时候要想好作品真正的价值是什么，自己真正在意的是什么。

I am inspired by the costume studies of China from 1950s to 1960s. There's a clear detail in my collection, that's the shirt collar. We can see from the Chinese historical documents from 1950s to 1960s that works and farmers at that time would tie a knot on the neck using a towel, which became the design inspiration of today's shirt collar. I hope that people will find some interesting details after wearing my clothes.

To be honest, we acquire inspiration from our life by chance instead of searching for it deliberately. I think there's no fixed mode of working methods. I always keep sensitive toward the things around me, and then add them into my collection. I think it is regretful if we always hear other people's stories instead of trying ourselves. Age is valuable for designers. For example, for many designers, they may create quite different collection when they are young and mature. With time going on, we begin to treat things with different perspectives, and our values will also change. When designing collection, we need to think about the real value of it, and what's the meaning of our creation.

王逢陈 / Wang Fengchen
中国 / China

我的设计灵感源于我自己个人身上发生的故事和围绕在我身边的人。大部分都是跟内心的一个波动相关。这次回到母校参展的三套衣服都是我在皇家艺术学院的作品。其中第一套讲述的是我个人的童年回忆。比较有意思的是，外表上看都比较简单，但在板型、廓型以及设计概念上还是比较深入的。我会考虑从样式上怎么去打造裤子和衣服之间的关系。你们看到我的 lock1 的设计，裤子和夹克是一体的，我用了很多一片布的裁法把它们整个连起来了。

既然要把这个东西做出来，从造型上、外观上、设计上都好看的话，其实背后要下很多功夫，但可能成品出来会感觉蛮简单的。你去深究、去解读它的话，还是很有意义的。

这个系列中有一件衣服，也是反映我个人感受的，一个人从小长到中年再到老年的成长过程。我的作品主要就是以人为中心。每一个人，发生在他们身上的故事都不一样，所以我特别关注一些在岁月上有所经历的人。这个灵感延伸到我的衣服上的时候，我就会从服装上的接片出发。表面看着特别平常，但事实上我觉得衣服是人体的第一层皮肤。我在这个缝与缝之间做了弹力拉伸的设计，这种缝会设计在人体的一些关键部位，使人体活动的时候保障一个很大的活动量，又起到了装饰的效果。

I am inspired by the stories happened around me and people surrounding me. Most inspirations are related to the fluctuation of the heart. All the three works are designed by me in Royal Academy of Arts. The first one is about my memory of my childhood. What's interesting is that it looks simple in appearance, but it's profound in terms of model, silhouette and design. I can think about how to deal with the relationship between pants and clothes. You can see from my lock1 design that the pants and clothe are integrated, and I use many pieces of cloth to connect them.

If you want your work to be beautiful in modeling, appearance and design, you should put in time and energy. But maybe the finished product appears to be quite simple. It is meaningful to research and interpret it.

There is a piece of clothing which can reflect my personal feeling, and the process from young to middle age until old. My works are human-centered. Everyone has his/her own story, so I pay attention to those persons who have experiences with time going on. Therefore, while this inspiration is extended to my clothes, I will start from splicing of the clothes. It seems quite normal, but in my heart, clothes are the second layer of human bodies. I use elastic stretch skill between seams, which are designed in key positions of human bodies. Therefore prodigious activity level is ensured while people doing activities, and can also achieve decorative effect.

王浩文 / Wang Haowen
中国 / China

大家好，我是王浩文。非常感谢北京服装学院能够邀请我来参加这次活动，这次我带来了三件作品。

其中有两件用于时装秀，还有一件用于做静态的展览，希望你们能够喜欢。这个系列的灵感来源是来自于我的胎记，和我自己的一些个人感受。带过来的这三套衣服的主要特色是它的面料和一些编织手法，还有一些肌理的处理，都是根据我个人情感的一些感受进行加工处理和开发的。整个系列都是以黑白灰为主，希望你们能够喜欢。关于这次活动我觉得做得非常成功、非常好，希望北京服装学院 55 周年生日快乐。

Hello, I'm Wang Haowen. I'd like to express my thankfulness to BICT for inviting me to attend this exhibition. And this time I bring three pieces of work, among which two are used for the fashion show, and another one is used for static exhibition. I hope you will like them. They are inspired by my birthmark and my own feelings. The main characteristic of these suits lies in their shell fabric, weaving skill and the management of some texture, these are all materials that are disposed and developed according to my own feelings. The whole series primarily use black and white, and I hope you will like them. I think this exhibition is quite a success and I want to say "Happy Birthday" to BICT for its 55th anniversary.

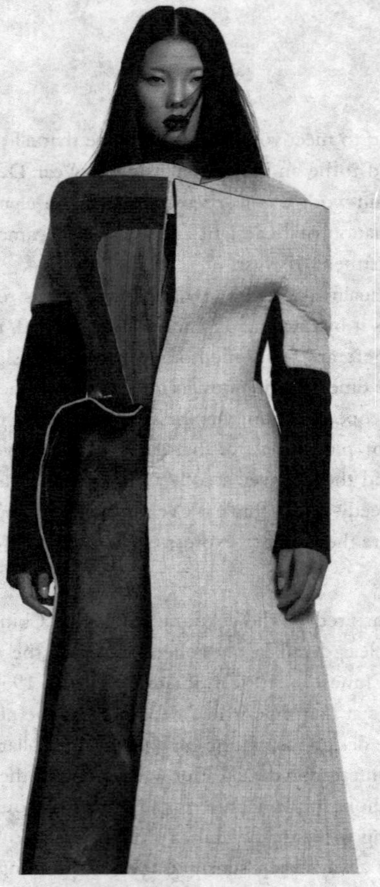

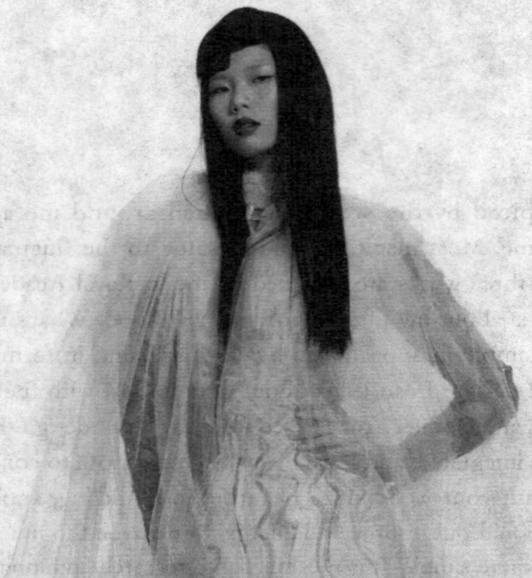

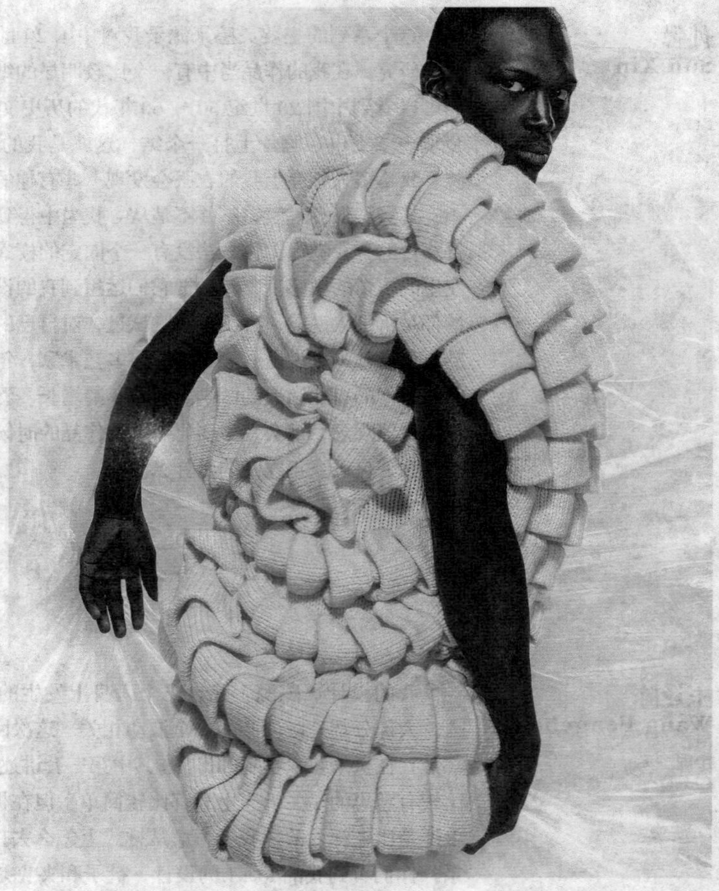

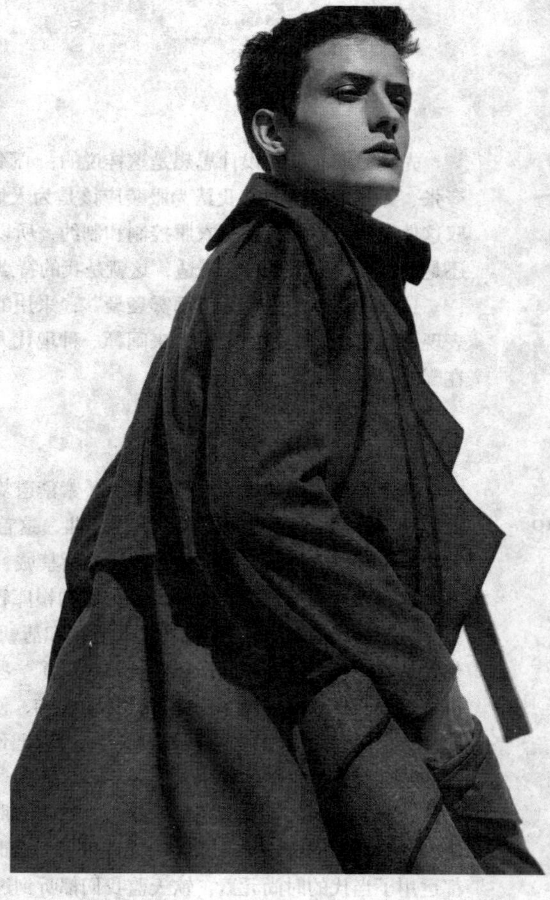

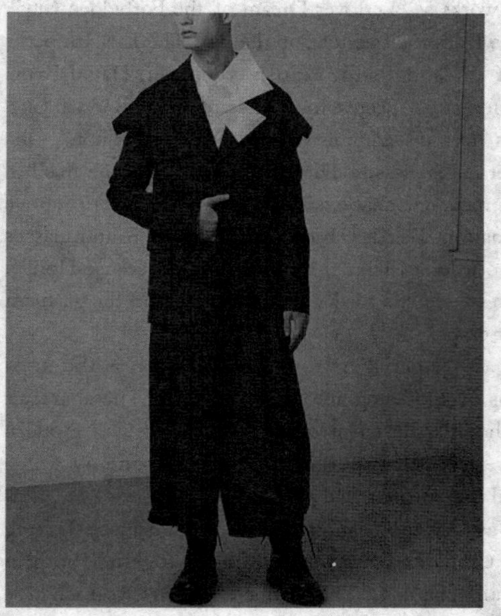

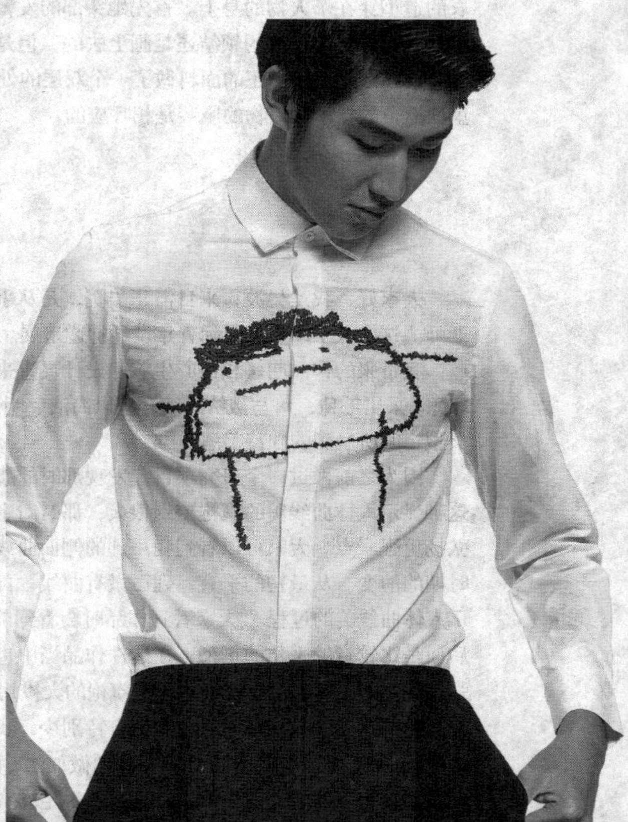

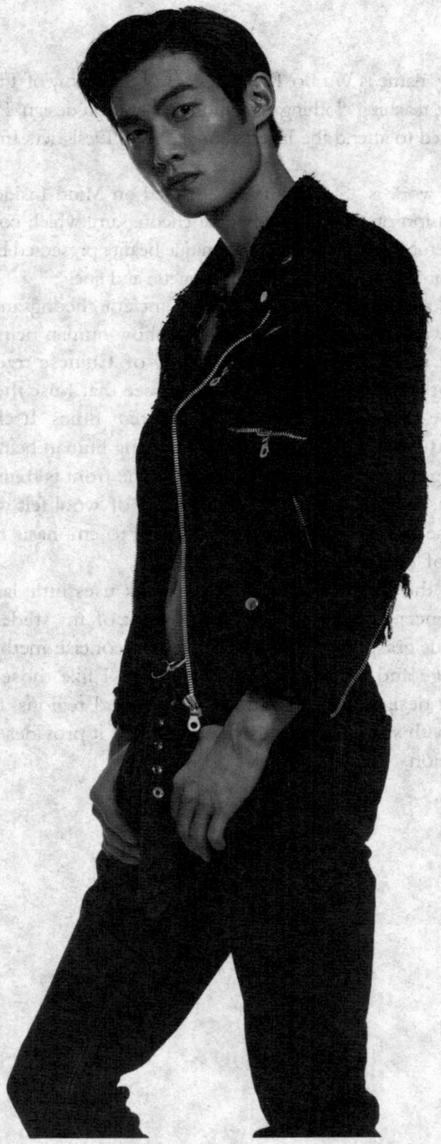

王珺 Wang Jun
中国 China

先秦时期有一句设计思想是这样说的：心不为物所易。我不喜欢太夸张、太笨重的服装，我认为服装应该是为人们服务的，穿着者应该驾驭这件衣服，不应该反被衣服控制和制约，所以我的作品无论是在面料还是造型上，都很柔，很舒适，这就是我的特点。

这次我作品的主题为"雨雾袅袅"，采用的面料是扎染的薄纱，想表现一种薄、透、轻的效果，来阐释一种现代人很舒适、很清逸，像云在飘的感觉。

There was a king of design philosophy during pre-Qin period: the heart can not be driven by materials. I don't like clothes which are too exaggerated and cumbersome. I think clothes are designed to serve the people, and those who wear them should have the ability to control them, instead of being controlled by them. Therefore, I design soft and comfort clothes whether in terms of materials or style, and this constitutes my design character.

The theme of my work is "Rain and Mist Curl Upwards". And I use chiffon as the material to create a king of light, penetrate and relaxed effect, illustrating a feeling that modern people are at ease, flowing above the cloud.

王文潇 Wang Wenxiao
中国 China

这个系列的服装我给它起名叫作《紫禁遗梦》，它的灵感来源是20世纪80年代的电影《末代皇帝》的序曲，这首曲子的作者是英国作曲家大卫·拜恩。外国人的视角看中国的紫禁城，可能会和我们不太一样。

这套衣服我给它起名叫作《弘仪阁的银库管理员》。其一是他拿的算盘，另外就是眼镜，看到算盘时，我觉得和清朝的齐头是有一些相似的，整体是这种长方形造型，所以我把算盘做了一些变形放在脑袋上。另外也给这个圆形的眼镜做了一些现代化的装饰，使它变成了一个像放大镜的样子。因为是有银行这样一个职能，所以把颈部包括衣服搭襟部分的一些装饰变成了铜钱的样子。

这套衣服叫作《钦天监的监正》。这套衣服是这个系列里面跟时尚造型结合最严密的一件，人物的面部带了一个面罩和肩部夸张的造型，都运用了当代的时尚元素，钦天监我们都听到过是管理天象的部门。

这套服装我给它起名为《兵工厂的兵器工程师》。清末时期，中西方不管是在文化还是科技上都有了相当程度的融合，衣服上这个人物造型我并没有给她穿中国传统的旗袍，而是把西方19世纪比较流行的风衣的造型穿了人物的身上。首先她头部的发髻刚好是做了这套衣服的肩部，另外这个人物的领结还是画上去的，但是到下面风衣打开的部分其实就已经是在用真正的面料做了一个双层的处理，外面是风衣的部分，里面蓝色的料子和人物的服装是相呼应的。

I name the series of works as "Leftover Dream of the Forbidden City", and the inspiration comes from the overture of the film The Last Emperor of 1980s. The writer of the song is the British composer, David Byne. Foreigners may have a different perspective to understand the Forbidden City.

I name the clothes as "treasury administrator of Hongyi Pavilion". One is his abacus and another is his glasses. The abacus looks like the flush in Qing Dynasty, and it's a rectangle shape, so I put the abacus form above the head after transformation. Besides, I also decorate the round glasses with modern elements, so it looks like a magnifying glass. Because Hongyi Pavilion had the function of bank, so I decorate the neck and the garment with the shape of copper cash.

I name the clothes as "Imperial Astronomers". And it is the most fashionable one in terms of modeling among all the suits. The character wears a mask and the shoulder design is quite exaggerated with modern elements. We all know that the imperial board of astronomy was a department that managed the astronomical phenomena.

I name the clothes as "Weapon Engineers of the Arsenal". During the late Qing Dynasty, China and the western mixed together both in culture and in science and technology. This character doesn't wear Chinese traditional cheongsam, but wears the wind coat which was quite popular in western countries in 19th century instead. Firstly, her bun serves as the shoulder part of the clothes. Besides, the bow tie is painted on. But in the unfold part of the wind coat, I use real material to create a double-layer effect. The outward appearance is the wind coat, and the blue material inside is connected with the clothes.

吴波 Wu Bo
中国 China

大家好，我是吴波，来自清华美院，是从事服装教育和服装设计的老师。这次很荣幸参加国际青年设计师邀请展。

我带来的作品叫《相由心生》。设计灵感来源于对中国传统旗袍的印象，采用毛毡、绡与戳绣的结合，以简洁、明快的廓型与线条强调正、负型之间所呈现的独特美感。

因为之前做过一个研究项目，发现那时的旗袍和我们现在所看到的这种显示人体曲线美的旗袍差别很大，那时对中国传统文化的含蓄美的弘扬更强一点。大家可以看到我用到的侧面曲线的形，这个形也有一定时期的演变。从最初的平直，到微微有曲线，再到通过一些手段达到显示人体曲线美的过程。大家看到作品时会看到前面是纱的透明的负型，后面是比较宽的毛毡的正型。在这件作品当中也用到戳绣的手法，一是强调线条的美感，二是在细节上把旗袍的大襟用这种手法强化出来。

我的这件作品相较而言用的语言特别少，我的设计理念就是"少即是多"。我的很多学生在设计中一味地做加法，而不知道去把握这个度，其实用较为简洁的方式就可以将语言说得足够有力了，传达的信息也是非常大的。很多设计师的作品我也喜欢，来自不同的民族和地区，风格迥异，地域性非常强，很有地方特色，这也是一个很好的交流平台。

Hello everyone, my name is Wu Bo, I'm a teacher of Academy of Fine Arts of Tsinghua University, teaching clothing education and fashion design. I feel it a great honor to be invited to attend the International Young Designers Invitation Exhibition.

The name of the work is "State Outside Is Based on Mind Inside". It is inspired by the impression of Chinese traditional cheongsam, which combines wool felt, raw silk and embroidery, stressing the unique beauty presented between positive and negative through simple and vivid silhouette and line.

I once conducted a research, and I found that ancient cheongsams were quite different from what we see today, which can show human being's line of beauty. In ancient times, the implication beauty of Chinese traditional culture was carried forward more strongly. You may see that I use the shape of profile curved line, which also changed in different times. It changed from being straight, to slightly curved, and until showing human being's line of beauty. After seeing the work you may find that in the front is transparent negative edge of yarn, and at the back is positive edge of wool felt which is a little wide. I also use the skill of poked embroidery to emphasis the line beauty and the front of cheongsam in detail.

You have caught the point. By contract, the work uses little language because my design concept is "less means more". A lot of my students use more language without grasping the degree. In fact, a concise method can replace much language and convey a lot of messages. I like those works designed by different designers from various nations and regions, they all have different styles with strong regionalism. And I think it provides a good stage for communication.

张晓田
Zhang Xiaotian
中国
China

我是2012年从北京服装学院毕业的，毕业后我去了纽约攻读硕士学位。我这次非常荣幸在北京服装学院建校55周年校庆的时候被母校邀请来参加这个国际性的设计师展览。首先要祝北京服装学院55周年生日快乐！学校现在的发展越来越好，我毕业仅仅才有两年的时间，回到这里，看到非常大的变化，包括一些学生的作品，还有学校整个的建设都是越来越好。

我这次回来参展的作品，是我在帕森斯设计学院的硕士毕业作品，也是刚刚在纽约时装周走秀过的。作品的灵感来源是大脑的神经细胞。我了解得很深入，翻阅了很多科研文献，查阅了研究人类大脑的一些不同项目，来寻求更多灵感。这一系列服装采用了大颗的珠子和不同的面料，都经过我亲自染色。糖果样的珠子不仅用作装饰，更起到去控制衣服结构和功能的作用。衣服的两端有或大或小的珠子，用来构建层次感，塑造身体的结构感。

我觉得设计师需要经历非常重要的一个环节，就是通过实验得到真正想要的、能真正做到的东西。当然你会遇到瓶颈，做了很多的调研，很多的样本，却不是自己想要的，感觉不是你想要的那个效果，这点确实很难。你需要努力挣扎、找到出路。而一旦找到出路，你会意识到这就是我想要做的，这就是我要专注的。我能做的事像是没有极限，如果真的努力、真正去尝试，我可以做任何事。

I graduated from BICT in 2012, and I went to London for master's degree after graduation. I feel it a great honor to be invited to attend the international designers' exhibition contest during the 55th anniversary of BICT. Firstly, I want to say happy birthday to BICT for its 55th anniversary! The school is developing in a fast speed. I just graduated for two years, but I see great changes, including students' collection as well as the construction of the school.

The collection that I bring this time is the master degree graduation work when I study in Parsons School of Design. And I just attended the New York Fashion Week with this collection. The inspiration mainly comes from nerve cells of the brain. In order to seek for inspirations, I made a deep research and referred a lot of scientific research literature, and looked up various items about human brains. This series uses big beads and different fabrics and I dyed in person. The candy-shaped beads are not only used for decoration, but also controlling the structure and function of clothes. At the both ends of the clothes, there are different sized beads, which are used to create a sense of layering and mould a sense of composition.

I think what a designer should experience is to gain what he/she really wants and can achieve through experiments. Of course you will meet bottlenecks, for example, you may not get what you want after many researches and samples, and it is quite difficult to deal with. You need to struggle and find the way out. And once you find the way out, you will realize that it is what you really want and what you should be focused on. There seems to be no limit for what I can do, and I can do everything if I make efforts and try.

杨大伟
David Yeung
中国（香港）
Hong Kong, China

我谈一谈我的想法。我的观点是西方对于中国的影响，我可能会用来自西方风格的绒毛，来展示现在西方风格对中国有很大影响。与此同时，我会在男性和女性的衣服上都使用圆形的纸筒，来表示中国文化的"虚弱"，由此可见西方文化影响已经扎根于中国文化当中，很多西方时尚在中国流行。很多国人去国外旅游，开拓了自己的视野，然后对于成衣，像在女性和男性服装上的圆形设计，来展示中国文化仍是根深蒂固的，仍有自己的风格。所以你可以很直观地看到这两种区别，我觉得这是很有趣的。

Ok, I'll talk about my opinion. My point of view focuses on the influences of western countries to China. I will use fluff with western style to illustrate its influence toward China. Meanwhile, I will use rounded fiber containers in men and women clothes to show the "weakness" of Chinese culture. It can be seen that the western cultural influences are anchored in Chinese culture, many western fashion styles are quite popular in China. A lot of people go abroad for travelling, and broadened their horizon, and then use circular design on women and men clothes to show that Chinese culture is deep-rooted with its own style. Therefore we can easily see the two distinctions and I think it's quite interesting.

Borre Akkersdijk
荷兰
Holland

一般来说，我作为染织设计师，我的工作是要把图案一点一点印到码布上。所以，我们在工作室里工作的时候，我们不用买现成的码布，而是真正自己制作码布。在全部的产品生产流程里，直到最终成型、送出展示之前，我们使用缝纫机和编织机这两种机器来工作。

当我们用码布、缝纫机、编织机工作的时候，从那些机器运行的原理中我们就能获得灵感。比如这些机器运行的状态、工作模式、印染方法、编织技术，以及作品最终完成的时候，我们有时候会发现一些令人惊喜、出人意料的事情。包括每次作品的图案、颜色、状态都对我们未来的设计有很大的影响。一句话，我们的灵感来自任何事物。

首先说说服装秀。服装秀举办得真的非常好，因为我可以看到很多新潮的设计理念，在后台了解了这场秀的流程以及其他设计师的作品。我也很开心看到这里的工作人员、模特能够在这个设计展上认真努力地工作，包括服装秀，在我看来非常专业、非常酷。我也发现了自己感兴趣的作品。如果可以的话，我很乐意经常来中国看看。

Generally speaking, as a dyeing and weaving designer, my work is to print the pattern onto the plaiting. Therefore, we don't have to buy ready-made plaiting, instead we make plaiting ourselves. We use sewing machine and knitting machine during the whole production process, until the final molding and before demonstration.

We may be inspired through the machine operating principle when we are using plaiting, sewing machine and knitting machine. Through the machine running status, work mode, printing and dyeing method, weaving skill as well as the completion, sometimes we may find things that are surprising. We are greatly influenced by the pattern, color and condition of each work. In a word, we are inspired by everything.

Firstly, I want to talk about the Fashion Show. The Fashion Show is excellent, we can see many fashion design concepts, know about the whole process behind the scenes and other works. I'm very happy to see that the works and models work hard. For me the design exhibition, including the fashion show, is quite professional and cool .I also find something that I'm interested in. If possible, I hope that I can come to China frequently in the future.

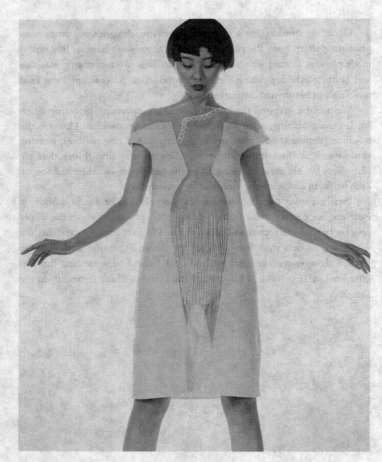

PT. Musa Atelier
印度尼西亚
Indonesia

我的作品灵感主要来自一个即将成为新娘的故事。新娘名为 Dara，这个故事我取自苏门答腊岛，在西苏门答腊的宗教信仰中，Dara 代表着新娘。因为我来自印度尼西亚，所以我需要表现出自己的特色。

我想要做的就是展示我自己的作品和在印度尼西亚售出自己的作品，所以我必须很了解印度尼西亚市场和它的流行趋势，然后很好地融合它们。有时候我会从传统故事中，某种织物中，某种手工艺者的赖以生存的活动中获得灵感。通过这种方法，我想让印度尼西亚文化走向世界。

I am inspired by a story about a girl who is going to be a bride. The bride is called Dara. I heard this story at Sumatra. According to the religion of Sumatra, Dara stands for Bride. Because I come from Indonesia, so I have to design something different.

What I want to do is to demonstrate my own works and sale my works in Indonesia. So, I must know much about the market and fashion trends of Indonesia, and then try to merge together. Sometimes I am inspired by traditional stories, fabrics and activities upon which crafters are relied. I desire to promote Indonesia Culture to go to the world by this way.

梅田悠希
Umeda Yuuki
日本
Japan

你好，我是日本文化学院大学服装设计学研究室的助手梅田悠希。

这次带来了七件作品，三件黑色的，三件红色的。这些都是我所教的学生设计并制作的，学生都是在大三的第二学期开始设计制作，在大四一开学进行展示。红色的作品是以亚洲民族服饰为主体设计的，而三件黑色的作品是根据环保、再利用为主题设计制作的。还有一件是我自己设计并制作的衣服，由于我现在正在研究美术作品与现代时尚的关系，所以我就以如何将美术作品转化为现在的时尚为主题设计了这件作品，我是将1910年的颜色与美术作品转变为现代时尚服饰来制作的。

Hello, I'm Umeda Yuuki, assistant of College Clothing Design Laboratory of Japan Cultural Institute.

I bring seven pieces this time, among which three are black and three are red. They are designed and made by my students from the second semester of junior, and began to demonstrate when they were seniors. The red ones regard Asian national dress as the main part, while those black ones are designed with the theme of environmental protection and recycling. The other one is designed and made by myself. Currently I'm studying the relationship between fine art and modern fashion, therefore I design this work with the theme of how to convert fine art into modern fashion. I made this work by using the color of 1910 and studying the convention of fine art to modern fashion clothes.

艾娜·贝克
Aina Beck
挪威
Norway

我认为每个人都有不同的风格与特性。我的作品以布料为主，注重材质。我想这也是作品的不同之处，因为是以实验为基础的。我的作品融入金属，有光泽。所以我使用纤维织物，十分具有个性。

我一直喜欢展示、绘画与设计。这也是我一直想做的。从我很小的时候就开始喜欢设计。我在挪威长大，然后去旅行，我去过西班牙。我在西班牙定居，之后又在英国定居，这个过程很愉快。之后我去了英国的艺术学校，后来又回到了挪威。我想去纽约，因为那里是时尚的天堂。我喜欢按照自己的意愿去设计。但是我十分注重纹理与面料。我喜欢做自己喜欢做的事情。

我对待工作十分认真，我喜欢自己的工作，很惊奇。我经常旅行，遇见形形色色的人，我也从中获得很多灵感。我喜欢中国的工作环境，也能从中获得灵感。

是的，我在为下一部作品搜集素材。我现在在画设计图，非常有灵感。这是我第一次来中国，甚至是第一次来亚洲，现在我十分喜欢这片国土。昨天我观察了两条街道，发现有太多的东西去探索。我看到了漂亮的衣服、漂亮的纹理。我希望能够在这里找到可以进行创作的素材。是的，我喜欢这里，希望以后还会再来。这里的人们都十分友好，简直太棒了，与挪威完全不一样。这里的文化、人与食物都很棒。

虽然我没有最喜欢的设计师，但是我很喜欢LANVIN，它十分简约，但这种简约透着魔力。我的梦想就是也能够设计出这样的作品。我认为每个设计师都希望设计出属于自己的品牌。我不喜欢上电视，我喜欢待在荧屏外面专心工作。我只是希望通过自己的努力让人们穿上自己设计的衣服，并感到很棒的感觉，这就足够了。我不喜欢戏剧，我比较喜欢安静。

I think we all have different styles, and different identities. And my stuff is very fabricbased, very textile development. And I think that's why it's different,because it's very experimental. I work with all the metallic, very shiny stuff. So everything that I do make my own fabrics. I think maybe that's why it's different. So everything becomes very personal.

I'd always love to illustration and draw, and design. That's what I always want to do. I grew up in Norway and then I traveled, and I did a lot illustration, and I went to Spain, I lived in Spain, and then I also lived in England, yeah is been nice and then I went to art school in England, and after that I went back to Norway, wanted to go to America to New York, it is good platform for fashion in New York. It's many opportunities and designer houses you want work for it. And design things that I always want to do. But I am very texture-based, very textiles, I'd like to do my own things.

I take my job very seriously, I love my jobs, I think it's amazing. And I get to travel, and meet lots of people, and that give me lots of inspiration. When I work its well to be here in China, and see the all material, that you have here, it's very inspired.

Yeah I am looking for fabrics, work on my next collection. Now I draw the designing, yeah I think so, I think it's very inspiring. This is my first time in China, never been before, and I never been to Asia before, but so far, and I really love that. I went to like two streets yesterday, it was so many things to look out. And I saw beautiful clothing, beautiful fabrics. And I really want to see if I can get fabrics here that I can work on. Yeah absolutely. I love be in here I definitely want to come back. Everyone is so nice here, very friendly, yeah it's greet. It's very different from Norway. You have been to Norway so you know. All the culture, the people, the food is fun.

I don't have favorite designers, but I do really like LANVIN, very simple, very magic simple. And my dream is to have my own. I think that for every designer to be able to design for their own label.I don't like to be on TV, I like to behind the scenes, and work, and be very dedicated to my work. I just want to be able to do my work, and the people to wear the clothing. Em… I just want people to wear my clothing and feel good. And I think that makes me happy. I 'd rather be in the background. I don't like any drama, nothing like that. I'm very quiet.

Sumy Kujon
秘鲁
Peru

我来北京参加这次盛会，希望能宣传秘鲁的文化。我设计的服装是用秘鲁的特色织物制成的，如羊驼毛、羊驼呢等织物。我设计的服装包括大衣、夹克、裙子、针织衫、饰品等。所以今天展示的服装也多种多样，丰富多彩。

与中国相比，中国从事服装设计的人更多，但我觉得拉丁美洲的潜力更大，因为我们有非常悠久的服装文化、工业文化等，还有我们特有的秘鲁纺织品。所以，我认为，有了这种独特的风格，不论在拉丁美洲还是在全世界，我们都将成为越来越主流的服装设计师。我个人非常喜欢纺织品。我觉得它给人感觉非常舒服。我始终着眼于自己国家的文化。我觉得这是我做得最正确的事。反映自己的文化，同时也跟进现代风格。这是一个融合的过程，很有意思。

今天是我第一次来到北京服装学院，我非常喜爱！开个玩笑，我甚至很嫉妒这里的学生，因为我的国家没有什么学校能比得上这所学校。我觉得非常受启发。

I come to Beijing to attend the grand meeting, with the purpose of advertising Peru culture. The clothes designed by me is made by fabrics with Peru characteristics, such as alpaca. I design overcoats, jackets, skirts, knitwear, accessories and so on. Therefore, I will show various clothes today.

Therefore, I will show various clothes with feminization today. There are more designers in China, but I think Latin America has more potential, because we have a long history of costume culture and industrial culture, especially Peru textile. Therefore, I think that with this unique style, we are doomed to be mainstream designers whether in Latin America or in the whole world. Personally, I prefer textile fabrics, which are comfortable. I always focus on the culture of my own country, and I think it is the right thing to do. Reflecting our own culture and following the modern style is a process of integration, so it is quite interesting.

This is my first time to visit your school and I like it very much. I am even jealous of you, of course just kidding, because our schools can not be compared with your school. And I am very much inspired.

艾琳娜·普拉托洛瓦
Elena Platonova
俄罗斯
Russia

我叫艾琳娜·普拉托洛瓦，我是俄罗斯人，但是我一直是在西班牙学习和工作。

我这次参展的作品叫作《泡泡》，它是一个气体，它也是一个概念，来源于一部科幻片的想法。我的这个概念就是想做对现实生活更有意义的设计。我观察到现在的人们已经很在意环境的状况，而且可能在未来会更加注重这方面的保护，因为环境越来越恶化，人们可能会希望身边有一个与外界隔离的透明空间，像一个小的气泡一样来保护自己，保护他们的孩子。当人们生活在这种状态下的时候，他们就会变得更加脆弱，当他们变得更加脆弱的时候，就会想要更好地保护自己。

在我的作品画集当中，是由两个小女孩来表现我作品的，用儿童来当模特，我其实想表达的就是他们代表的是一种纯洁，是一种非常简单，没有被污染的一个生命体。我的作品一直都想表现在这种很简洁、很干净的状态下，人们的生活方式。我的作品展现出来的面料是柔软的，色彩是非常清淡洁白的，还有就是线条很简单，甚至无细节。主要想表达的就是人们对自身的保护状态，保护意识。

这个展览给了我们一个展示自己的机会。同时，对于我来说，也非常高兴能够了解中国或者是亚洲这个市场，或者是整个这个设计领域现在是一个什么状态。因为欧洲和亚洲在服装方面其实还是有很大的风格上的差异，但是我非常高兴能够亲身体验到或者亲身感受到这种差异。

My name is Elena Platonova. I come from Russian, but I am studying and working in Spin all the time.

The name of the work is Bubbles. It is a kind of gas, but also a conception, which drives from an idea of science fiction film. I design the work with the purpose of doing meaningful things to real life. I notice that people are paying more and more attention to the environmental conditions, and maybe they will be more aware of self-protection in the future. Due to the deterioration of environment, maybe people are hoping to have a transparent space which is isolated from the outside world to protect them as well as their children like a bubble. People are likely to become weaker under this kind of condition, so they will have stronger desire to protect themselves than before.

I select two little girls as models to indicate that they represent purity, and they are life entities that are very simple and without being contaminated. I always want to illustrate people's life style, self-protection condition and awareness under the simple and clean situation. My design works have three characteristics: firstly, the materials are soft; secondly, the color is quite light; thirdly, the lines are very simple. What I want to express is people's self-protection condition and awareness.

This exhibition gave me a change to show myself. Meanwhile, I'm very glad that I can understand China or the market of Asia, and I also get to know the design condition through this exhibition. Though there are many differences in style between Europe and Asia, I feel it a great honor to experience these differences personally.

Jon Mikeoezkurdia
西班牙
Spain

我带来的设计是我的毕业设计之一。它所讲述的是我家乡的传统文化，我的设计中有非常多的制作蕾丝的工具，一种碰在一起会轻敲出声的小木棍。它所代表的西班牙20世纪50年代一些古老服饰的传统特点，我试图去重新塑造这种传统，并且让它变得更概念化，利用那些声音，还有新颖的材料。

羊毛是这件设计的面料，但我想让它变得更独特新颖，所以我决定添加蕾丝小工具这个元素，也就是木制品。所以最后的效果就像是我的家乡，西班牙巴塞罗那的一种乐器的声音。我当时就很肯定羊毛和木头的结合碰撞会非常有趣。也因为这样，当在T台上，你是可以真实地听到我所创造的声音的。

如果是年轻的设计师的话，你需要被大众发现，如果你希望被发现，你就要很疯狂。比如说，如果你只是设计一件T恤，大家根本不会意识到你的存在，因为太多的设计师，太多的品牌在做。如果你想变得与众不同，首先你就要做和别人不一样的事情，但是之后如果你不做一些可穿性高的设计，你就没有前途。所以我认为每一个时尚行业的人或者至少是在学习时尚的人，最开始的时候，你都必须要表达出最新潮的想法。

我确实认为，美国式的时尚和西班牙式的时尚不太一样。伦敦的时装设计可能更具实验性，更新潮，是很独特的潮流。巴黎，你会看到很棒的高级定制，那些最好的、最有名的品牌都在其中。而在意大利，你可以见到最好的匠人，最好的剪裁。每个国家都有完全不一样的时尚，当然西班牙也一样。西班牙也有和其他国家不同的品牌，比如罗意威（LOEWE），而且他们其实都很棒。但是，是的，要成为一名时装设计师，西班牙确实不是世界上最好的地方

The design I brought is one of my graduation designs. It is about the traditional culture of my hometown. In my design, there are many tools used to make laces-a kind of club stick that will generate sounds when crashing each other. Therefore, it stands for some traditional characteristics of ancient costumes of Spain in 1950s. I try to remould this tradition, and conceptualize it, using those sounds and novel materials.

The design material is wool, but I want to make it more unique and novel, therefore I decide to add the element of lace tool-woodwork. Therefore, the final effect sound is like a kind of musical instrument sound of my hometown-Barcelona. I was sure at that time that the combination of wool and wood would be very interest. And so you can hear the real sound on the T stage.

If you are a young designer, you need to be discovered. And if you want to be discovered, you should be very crazy. For example, if you just design a T-shirt, then people will pay no attention to you. Because many designers and brands are designing these things. If you want to be unique, you should do different things firstly. But if you don't design things with high wearable, you are impossible to be succeed. Therefore, I think those who belong to fashion industry or are learning fashion should express the latest ideas at the very beginning.

I do think that American fashion and Spanish fashion style are different. The fashion design of London is more practical and novel as well as unique. In Pairs, we can find great advanced customization and many famous brands. While in Italy, you can see the greatest craftsmen and the best clipping skill. Every country has its unique fashion, and Span is not an exception. Span also has its own brands, such as LOEWE, and they are quite great. But in terms of being a fashion designer, Span is not the best place.

吴亭筠
Wu Tingjun
中国（台湾）
Taiwan, China

大家好，我是来自中国台湾的吴亭筠，我就读于台湾实践大学，今年刚从实践大学的服装设计研究所的服装设计专业毕业。

这个作品的名字叫《绣》。借由绣，揭露我自己这个阶段的一个处境或者是一个情况，我觉得绣非常有意思的地方是，它是一种覆盖动作，在覆盖动作之上其实它会同时揭露出很多关于绣者的一些情感跟它当下的处境或者状况，从而会呈现出不一样的针法，众多的针法、情感等加织、交缠在上面，所以我在编号我的每一件成品的时候都会想起我在尝试和创作过程中一些细枝末节的心情……

我的创作出发点是想重新探讨或重新定义衣服跟人的关系，或者说人跟物的关系。所以我一直在重复使用车机在进行绣这个动作。我希望大家看到的、大家想要知道的是我在做这件作品的时候，它背后的意义，也希望大家重新思考在这样的制作方式下它能够带来什么样的反思。

我的制作过程就像我刚刚讲述的一样，它其实就是一直在用工业车机进行绣的动作，它是乱针刺绣，它跟织一块布或者是在一块布上进行更多的刺绣比较有关系，我织了超过十块的布，我做了一些小机关，让那些布能够披挂在人身上或者是能够让人穿搭得很好看。如果大家有兴趣的话可以过来问我关于这件作品的制作过程或者它背后更多的故事。

Hello everyone, I'm Wu Tingjun from Taipei, China. I studied in Shih Chien University, majoring in fashion design of Fashion Design Institute, and graduated this year.

The name of the work is "Embroidery". I want to disclose my condition in this stage through this work. What interests me most is that embroidery is a kind of override action, which can disclose the emotion or condition of the embroiderer. And therefore, various stitches and emotions are reflected, and fold on the work. So, while numbering my finished works, I would think of the emotion of trying and creating….

I want to reexplore or redefine the relationship between clothes and human beings, or the relationship between human beings and objects. Therefore, I am repeating the action that machines are embroidering. I hope that people can figure out the real meaning of my work, and that people can rethink what reflections it will be brought through this kind of production method.

Just like what I mentioned above, it is using industrial machine to embroider and belongs to disorderly embroidery. It is related to weaving a piece of cloth or conducting more embroidery in one piece of cloth. I weaved more than ten pieces of cloth, and tried to make it matching while people wearing it. If you are interested, you may come here to ask me the working process or the background of this work.

许乃仁
Xu Nairen
中国（台湾）
Taiwan, China

我这次带来的作品的设计主题是《成人礼》。我大三的时候正好是20岁，在亚洲地区都会有行冠礼，行冠礼就代表这个男子正式成年。社会会赋予他一个象征身份的意义。

我用了一些绑带式的东西，中国的冠礼会在头上做一个绑系，因为冠礼就是行帽之礼，但是我觉得用冠会太传统，我想要简化一点线条的方式去做那种绑缚在身上的形象。所以我才会做一些绑带和流苏这样很祭典式的东西放在身上。

颜色方面我选择了我很喜欢的颜色——桃红色作为我这次作品的主要色彩。我用了一些像是补花的方式和一些针织面料的图案去表达，有点斑驳，就像是一层层快要脱落的身体在蜕化中的感觉。

我一贯的设计方式就是我喜欢从我自身的故事去思考，再想当时的自己跟衣服有什么关联性，再去寻找衣服的可能性。我比较喜欢抽象的情感，所以也常常找不到方向，但是我觉得那就是一种乐趣，就是去寻找自己在当下最喜欢的东西，这次从打板打样到制作全部都是自己来，完成了六套作品。

The design theme of my work is "Adult Ceremony". I was 20 years old and a junior at that time. In Asian regions, there are capping ceremonies, which represent that the boy becomes an adult. And the society will endow him a sense which represents his status.

I use something with bind belt types. There is a tie on the capping ceremony in China. Because capping ceremony is cap saluting ceremony, so I think it is much too traditional to use crown. I want to express the images that bind human being's bodies by using simple lines. Therefore I add festival things like bandage and tassels.

I choose colors which I like, and I select pink as the main color this time. I use appliqué method and knitted fabric patterns to express, which looks like mottled and the body that is going to peel off to degenerate as the design element.

My usual design method is that I like to reflect based on my own story, and then think about the relationship between myself at that time and the clothes, and then search for the possibility. I prefer abstract feelings, so I also loss myself. But I think it's fun to search what we like most at the time being. This time I completed the whole process from pattern making, to sampling making and until producing. And I made six sets in total.

附录
设计师
DESIGNERS

造型艺术	艺术设计	服装设计
潘松 / 中国	Tanel Veenre / 爱沙尼亚	Chris Ran Lin / 澳大利亚
杨正 / 中国	Simon Cottrel / 澳大利亚	Elena Platonova / 俄罗斯
于元尚 / 中国	Tabea Reulecke / 德国	Borre Akkersdijk / 荷兰
官卫康 / 中国	Michel Belin–Benhamou / 法国	Vladimira B.Steffek / 加拿大
	Cody Hudson / 美国	Kim Shui / 美国
	Bud Rodecker / 美国	Sumy Kujon / 秘鲁
	Craighton Berman / 美国	Aina Beck / 挪威
	Mikal Hallstrup / 美国	梅田悠希 / 日本
	Marta Mattsson / 瑞典	日本文化学园大学 / 日本
	李恒 / 中国（台湾）	许乃仁 / 中国（台湾）
	Danielle Cohen / 以色列	吴亭筠 / 中国（台湾）
	周依 / 意大利	Jon Mikeo Ezkurdia / 西班牙
	Thomas Hicks / 英国	David Yeung / 中国（香港）
	Naren Barfield / 英国	PT. Musa Atelier / 印度尼西亚
	Raz Barfield / 英国	暨剑侠 / 中国
	Jose Romussi / 智利	Mengjie Di / 中国
	隋宜达 / 中国	Xiaotian Zhang / 中国
	孙捷 / 中国	郭画 / 中国
	覃京燕 / 中国	王逢陈 / 中国
	胡桉澍 / 中国	王浩文 / 中国
	梁鹂 / 中国	王珺 / 中国
	贺阳 / 中国	王文潇 / 中国
	庄冬冬 / 中国	吴波 / 中国
	曹毕飞 / 中国	孙馨 / 中国
	张清绚 / 中国	白鸽 / 中国
	黄河 / 中国	倪苀 / 中国
	呼乃纾 / 中国	黄衍之 / 中国
	何为 / 中国	
	唐天 / 中国	
	罗骁 / 中国	
	周晓童 / 中国	
	崔赢 / 中国	
	叶子乐 / 中国	
	蔡志勇 / 中国	
	周易 / 中国	
	张恩倩 / 中国	
	张亮 / 中国	
	贾大鹏 / 中国	

内 容 提 要

2014北京服装学院国际青年设计师邀请展分为服装服饰动态展区及设计与艺术展区。本作品集收录了此次展览中来自世界21个国家和地区的70余位青年设计师的精美作品，充分展现了他们的设计才华。

本作品集分为四部分。第一部分为服装设计作品，这些优秀作品从多角度展示了国际青年设计师的服装设计创新能力，从中可以感受到来自不同文化传统的国度对服装服饰设计的理解与诠释。第二部分为艺术设计作品，包括雕塑、绘画、装置、陶瓷、首饰、服饰、创新面料、影像、动画、摄影作品和互动服装服饰产品。这些作品呈现了时尚语境下青年设计师对于生活方式的塑造和对于美的追求。第三部分为造型艺术作品，通过对作品造型的诠释，表现青年设计师的艺术才华。第四部分为创意说，通过与青年设计师的对话，让读者更加了解青年设计师对设计、对未来、对当下的认知。

图书在版编目（CIP）数据

未来即现在：2014国际青年设计师邀请展作品集 /
刘元风主编. —北京：中国纺织出版社，2015.10
ISBN 978-7-5180-1885-7

Ⅰ. ①未… Ⅱ. ①刘… Ⅲ. ①服装设计—作品集—中国—现代 Ⅳ. ①TS941.2

中国版本图书馆CIP数据核字（2015）第183082号

责任编辑：张思思　　责任校对：余静雯
版式设计：李　煌　　责任印制：储志伟

中国纺织出版社出版发行
地址：北京市朝阳区百子湾东里A407号楼　邮政编码：100124
销售电话：010—67004422　传真：010—87155801
http://www.c-textilep.com
E-mail:faxing@c-textilep.com
中国纺织出版社天猫旗舰店
官方微博http://weibo.com/2119887771
北京奇良海德印刷股份有限公司印刷　各地新华书店经销
2015年10月第1版第1次印刷
开本：787×1092　1/8　印张：18.75
字数：184千字　定价：198.00元

凡购本书，如有缺页、倒页、脱页，由本社图书营销中心调换